Ramesh Kandala

Three Body Scattering Problem

AF790953

Ramesh Kandala

Three Body Scattering Problem

LAP LAMBERT Academic Publishing

Imprint
Any brand names and product names mentioned in this book are subject to trademark, brand or patent protection and are trademarks or registered trademarks of their respective holders. The use of brand names, product names, common names, trade names, product descriptions etc. even without a particular marking in this work is in no way to be construed to mean that such names may be regarded as unrestricted in respect of trademark and brand protection legislation and could thus be used by anyone.

Cover image: www.ingimage.com

Publisher:
LAP LAMBERT Academic Publishing
is a trademark of
Dodo Books Indian Ocean Ltd. and OmniScriptum S.R.L publishing group

120 High Road, East Finchley, London, N2 9ED, United Kingdom
Str. Armeneasca 28/1, office 1, Chisinau MD-2012, Republic of Moldova, Europe
Managing Directors: Ieva Konstantinova, Victoria Ursu
info@omniscriptum.com

Printed at: see last page
ISBN: 978-620-8-11776-4

Copyright © Ramesh Kandala
Copyright © 2025 Dodo Books Indian Ocean Ltd. and OmniScriptum S.R.L publishing group

to my

Father late KANDALA JAGGAIAH

with love

Acknowledgements

It is my pleasure to thank the many beloved people who made this thesis possible. First and foremost I would like to express my heart-felt gratitude to my beloved supervisor, Dr. Najamul Hasan, Head, Department of Astronomy, Osmania University for his advice, encouragement, enduring patience and constant support. His never ceasing belief in me (though I had often doubts on my own abilities) gave me required belief in myself. Some of his other outstanding positive attitudes include clarity in explanations (when I was hopelessly lost), constantly driving me with energy (*Where does he get it?!*) when I was tired, and always, *always* giving me his time, in spite of anything else that was going on. His understanding, encouragement and personal guidance have provided a good basis for the present thesis.

My sincere thanks to, Prof. M.V. Ramana Murthy, Head, Department of Mathematics, Prof. M. Rangamma, Chairman, Board of Studies, Prof. B. Shankar, Prof. L. Anand Babu (Retd), Prof. J. Sucharitha, Prof. Mohd. Moiz Uddin, Dr. Alka Mashalker, Dr. J. Anand Rao, Dr. N. Kishan, Dr. M. Chenna Krishna Reddy, Dr. B. Krishna Reddy, Dr. C. Goverdhan, Dr. G. Kamala, Mr. S. Hari Singh Naik, Dr. B. Surender Reddy (University College of Science, Saifabad) and all the other faculty members of the Department of Mathematics. Their ideals and concepts created a remarkable influence on my entire career in the field of research.

My special thanks goes to Prof. G. M. Ballabh for giving his inestimable guidance in understanding the Sun-Earth-Moon system. I thanks to Prof. P. Vivekananda Rao, Prof. G. Yellaiah, D. Shanthi Priya and the Staff members, Department of Astronomy for all the guidance and support.
I'm thankful to the Librarian, Non-teaching Staff, Department of Mathematics and Department of Astronomy for their constant support.

During my research work, I worked with many colleagues for whom I have great regard, and I wish to extend my warmest thanks to Dr. K. Chenna Reddy for all the guidance and support while writing thesis in LaTeX, Dr. Y. Ravi Kiron and Dr. A. Lakshmi and all those who have helped me at the Department of Mathematics and Department of Astronomy. I owe collectively and individually to my colleagues Dr. B. Narender, Ajay Kumar Chandrakor, G. Harshavardan Reddy, A. Sandhya Rani, B. Aparna and K. Santhosh for their continuous support during these five years. I wish to thank my friends Dr. V. Srinivasa Rao, V. Venu, P. Vijaypal Reddy and Dr. N. Raji Reddy for helping me to get through the difficult times, and for all the emotional support, entertainment and care they provided.

I'm grateful to Prof. K. Rama Reddy, Senior Director, C. V. S. R. College of Engineering and Technology, Hyderabad and Prof. V.V. Haragopal, Department of Statistics, Osmania University, who motivated to purse research work. They introduced me to Dr. S. N. Hasan and supported in completion of my research work.

I wish to thank my parents Smt. Kousalya and *late* Sri. Jaggaiah for their love and affection and in-laws Smt. Bhagya Lakshmi and Sri. Murali for supporting in all my pursuits and providing a loving environment. It is a pleasure to express my gratitude whole heartedly to my brothers, Thirupathi, Srinivas, sister, Smt. Kavitha and Brother-in-laws Cheralu, Srinivas and Sadanandam who gave me strength to reach my goal.

And finally, I must thank my dear wife for putting up with me during the development of this work and still continuing, loving and supporting me without any complaints. I do not have words to express my feelings here, I can only say that I love you, Lakshmi!. Last but not least, lots of love to my kids Vikky and Lakky for taking me back to my childhood days and providing me instant energy to start over again with my work.

I'm sure I have missed out some of people who helped me directly or indirectly. Please pardon me with great heart if I have left you out.

Table of Contents

Page

Acknowledgements . . . ii

Table of Contents . . . v

Chapter

1 Three-Body Problem . . . 1

1.1 Introduction . . . 1

1.2 Mathematical description of the Three-Body Problem . . . 4

1.3 The Integral of Angular momentum . . . 6

1.4 The Integral of Energy . . . 7

1.5 Stationary Solutions of the Three-Body Problem . . . 10

2 The Dynamics of the Sun-Earth-Moon . . . 21

2.1 Introduction . . . 21

2.2 Sun's Perturbations on the Moon's Orbit . . . 23

2.3 Periodic fluctuations in moon's synodic period and advance of Moon's perigee . . 28

2.4 Conclusions . . . 35

3 Three-Body-Scattering Problem . . . 37

3.1 Introduction . . . 37

3.2 Classification of the motion of the three bodies in a plane . . . 39

3.3 Study of motion of three-bodies with numerical results 41
3.4 Historical and scientific background 42
3.5 Concept of solution 43
3.6 Solution 45
3.7 Conclusion 49
4 Scattering phenomenon in binary systems 57
4.1 Introduction 57
4.2 The effect of parabolic encounter on the angular momentum and eccentricity in a circular binary 58
4.2.1 Perturbation theory 58
4.2.2 Analytic expressions 62
4.2.3 Analytical discussion 66
4.3 The effect of parabolic encounter on the eccentricity of binaries 71
4.3.1 General expressions for the change in eccentricity 71
4.3.2 Binaries with non-zero initial eccentricity 74
4.3.3 Initially circular binaries 75
4.3.4 Numerical illustrations 78
4.4 Conclusions 81
5 The Scattering of Oort's cloud objects due to encounter with Giant Molecular Clouds . 84
5.1 Introduction 84
5.2 Giant Molecular Cloud 86

5.3 Equations of motion 89

5.4 Description of the simulations 92

5.5 Conclusions 94

6 Scattering phenomenon in a star cluster due to an encounter with interstellar cloud . . . 95

6.1 Introduction 95

6.2 Star - Star Encounter 95

6.3 Tidal effects of passing interstellar clouds on star clusters 100

6.4 Conclusions 111

7 Conclusions 112

7.1 Our work 112

7.2 Future Work 117

References 118

7.3 Publications 125

List of symbols

T	:Kinetic energy(K.E)
V	:Potential energy(P.E)
E	:Energy
L	:Angular momentum
$\dot{v}$	:Acceleration
u_r	: Radial unit vector
u_θ	: Transverse unit vector
G	: Gravitational constant
$M_\odot$	: Mass of the Sun
$M_\oplus$	: Mass of the Earth
m	: Mass of the Moon
a	: Semi-major axis
e	: Eccentricity
i	: Inclination between two orbital planes
Ω	: Longitude of ascending node

ω	: Argument of periapsis
n	: Mean motion
M	: Mean anomaly
$\psi^{'}$	: Mean longitude of the Sun
R	: Perturbation function
P	: Period of the Moon's orbit
ΔP	: Moon's amplitude of fluctuation
ΔE	: Change in energy
δe	: Change in eccentricity
p	: Cloud-centric distance
R_1	: Maximum radius of the GMC
ρ_1	: Density at R_1
N(e)	: Distribution of eccentricity
N(i)	: Distribution of galactic inclination
$p_{\odot}$	: Cloud-centric radius of the Sun
$\dot{p}_{\odot}$	: Cloud-centric velocity of the Sun

Chapter 1

Three-Body Problem

1.1 Introduction

The classical Newtonian three-body gravitational problem is one of the fundamental problems in Celestial Mechanics. It has the subject of investigations for over four hundred years. It has attracted attention of many great mathematicians, to name few Newton, Lagrange, Laplace, Jacobi etc.. For review of earlier works refer Gutzwiller (1998). Apart from the classical three body problem, restricted three body problem too has been of interests as it finds applications in satellite dynamics. The general three-body problem remains unsolved today but important advances and insights have been enabled by the advent of computers and modern computational techniques.

The long term stability of the orbits of the planets in our solar system and in particular the orbits of the Earth and the Moon was one of the early concerns. Newton solved the two body problem for the orbit of the Moon around the Earth and considered the effects of the Sun on this motion. This is perhaps the earliest appearance of the three-body problem. The first and simplest periodic exact solution to the three-body problem is the motion on collinear ellipses found by Euler (1767). Also Euler (1772) studied the motion of the Moon assuming that the Earth and the Sun orbited each other on circular orbits and that the Moon was massless. This approach is known as the restricted three body problem. Around the same time Lagrange (1772) discovered the equilateral triangle solution. The collinear and equilateral triangle solutions are the only explicit solutions known for arbitrary masses. The basis for the modern theory of the restricted three body problem was developed by Jacobi (1836), Delaunay (1860), and Hill (1878). The classical period ends with the powerful methods of surfaces of section, phase space and deterministic chaos developed by Poincare who was awarded the prize established by Sweden's King Oscar II to solve the n-body problem in 1889.

The most fascinated problem in Dynamical system is known as the problem of three bodies, which may be stated in the following way: "*The study of the motion of three mass points under mutual Newtonian gravitational attraction, so that between each pair of masses there is an attractive force which is directly proportional to the product of the masses of the particles and the inverse square of their distance*" is called the three-body problem.

Examples:

(1). The Sun - Earth - Moon system

(2). The Earth - Moon - Satellite system

(3). The Sun - Earth - Comet system

The geometrical structure of the three-body problem is as shown in the Figure - 1.1.

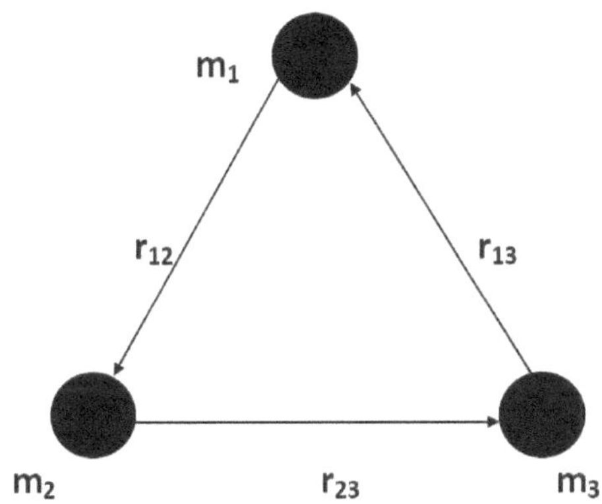

Figure 1.1: The three - body system

The significance of this problem arises from its applications to Celestial Mechanics: the bodies are considered to be the part of the solar system. They attract each other according to the Newtonian law of gravitation, in which the celestial bodies are replaced by particles of masses equal to the masses of the respective bodies. In this chapter, we present the description of three-body problem, the known integrals of the system and the reduction of the problem to a dynamical problem with a lesser number of degrees of freedom.

1.2 Mathematical description of the Three-Body Problem

The differential equations of motion.

Let m_1, m_2 and m_3 be the three arbitrary masses, let $\vec{r}_1$, $\vec{r}_2$ and $\vec{r}_3$ be the position vectors of three mass points with respect to an arbitrary origin and their mutual distances are r_{12}, r_{13} and r_{23} in an inertial coordinate system. Where we use the notation $\vec{r}_{ij} = \vec{r}_i - \vec{r}_j$ and $r_{ij} = |\vec{r}_{ij}|$ for all i, j. Let (x_1, y_1, z_1), (x_2, y_2, z_2) and (x_3, y_3, z_3) be the coordinates of m_1, m_2 and m_3 respectively. The equations of motion of three mass points are

$$m_1\ddot{\vec{r}}_1 = G\left\{-\frac{m_1m_2}{r_{12}^3}\vec{r}_{12} - \frac{m_1m_3}{r_{13}^3}\vec{r}_{13}\right\} \tag{1.1}$$

$$m_2\ddot{\vec{r}}_2 = G\left\{-\frac{m_2m_3}{r_{23}^3}\vec{r}_{23} - \frac{m_2m_1}{r_{21}^3}\vec{r}_{21}\right\} \tag{1.2}$$

$$m_3\ddot{\vec{r}}_3 = G\left\{-\frac{m_3m_2}{r_{32}^3}\vec{r}_{32} - \frac{m_3m_1}{r_{31}^3}\vec{r}_{31}\right\} \tag{1.3}$$

where G is gravitational constant.

This system consists of nine differential equations, each of second order, and hence the order of system is 18. No forces act on the particles except the mutual gravitational attraction, it can be shown that the center of gravity of the system moves in a straight line with uniform velocity. This can be expressed by the six integrals in the following way.

By adding the equations (1.1), (1.2) and (1.3), we obtain

$$m_1\ddot{\vec{r}}_1 + m_2\ddot{\vec{r}}_2 + m_3\ddot{\vec{r}}_3 = 0 \tag{1.4}$$

Integrating the equation (1.4) once with respect to time, we obtain

$$m_1\dot{x}_1 + m_2\dot{x}_2 + m_3\dot{x}_3 = c_1 \tag{1.5}$$

$$m_1\dot{y}_1 + m_2\dot{y}_2 + m_3\dot{y}_3 = c_2 \tag{1.6}$$

$$m_1\dot{z}_1 + m_2\dot{z}_2 + m_3\dot{z}_3 = c_3 \tag{1.7}$$

Integrating again with respect to time

$$m_1x_1 + m_2x_2 + m_3x_3 - c_1t = c_4 \tag{1.8}$$

$$m_1y_1 + m_2y_2 + m_3y_3 - c_2t = c_5 \tag{1.9}$$

$$m_1z_1 + m_2z_2 + m_3z_3 - c_3t = c_6 \tag{1.10}$$

Where c_1, c_2, ... c_6 are constants. By using these integrals, we can reduce the order of the equations of motion from $18th$ to $12th$ order.

1.3 The Integral of Angular momentum

Let $\hat{r}_{12}$, $\hat{r}_{13}$ and $\hat{r}_{23}$ be unit vectors in the directions of $\vec{r}_{12}$, $\vec{r}_{13}$ and $\vec{r}_{23}$. Then the equations (1.1), (1.2) and (1.3) can be written

$$m_1\dot{v}_1 = G\left\{-\frac{m_1 m_2}{r_{12}^2}\hat{r}_{12} - \frac{m_1 m_3}{r_{13}^2}\hat{r}_{13}\right\} \tag{1.11}$$

$$m_2\dot{v}_2 = G\left\{-\frac{m_2 m_3}{r_{23}^2}\hat{r}_{23} - \frac{m_2 m_1}{r_{21}^2}\hat{r}_{21}\right\} \tag{1.12}$$

$$m_3\dot{v}_3 = G\left\{-\frac{m_3 m_2}{r_{32}^2}\hat{r}_{32} - \frac{m_3 m_1}{r_{31}^2}\hat{r}_{31}\right\} \tag{1.13}$$

By taking the vector product of equations (1.11) to (1.13) respectively by $\vec{r_1}$, $\vec{r_2}$, $\vec{r_3}$ and adding, we obtain the following expression,

$$\sum_{i=1}^{3}\vec{r_i}\times m_i\dot{v}_i = G\left\{\frac{m_1 m_2}{r_{12}^2}(\vec{r_1}-\vec{r_2})\times\hat{r}_{12} + \frac{m_2 m_3}{r_{23}^2}(\vec{r_2}-\vec{r_3})\times\hat{r}_{23} + \frac{m_3 m_1}{r_{13}^2}(\vec{r_1}-\vec{r_3})\times\hat{r}_{13}\right\} \tag{1.14}$$

Since $(\vec{r_1}-\vec{r_2})$ is collinear with $\hat{r}_{12}$, $(\vec{r_2}-\vec{r_3})$ is collinear with $\hat{r}_{23}$ and $(\vec{r_1}-\vec{r_3})$ is collinear with $\hat{r}_{13}$. Therefore the vector products and, hence the R.H.S of equation (1.14) becomes zero.

$$i.e., \sum_{i=1}^{3}\vec{r_i}\times m_i\dot{v}_i = 0 \tag{1.15}$$

Therefore the above equation can be written as

$$\frac{d}{dt}\left\{\sum_{i=1}^{3}\vec{r_i}\times m_i v_i\right\}=0 \tag{1.16}$$

The quantity in the bracket represents the total angular momentum of the system of three mass points around each of the coordinate axes is constant throughout the motion.

The equation (1.16) can be expressed by the following equations.

$$m_1(x_1\dot{y}_1-y_1\dot{x}_1)+m_2(x_2\dot{y}_2-y_2\dot{x}_2)+m_3(x_3\dot{y}_3-y_3\dot{x}_3)=c_7 \tag{1.17}$$

$$m_1(y_1\dot{z}_1-z_1\dot{y}_1)+m_2(y_2\dot{z}_2-z_2\dot{y}_2)+m_3(y_3\dot{z}_3-z_3\dot{y}_3)=c_8 \tag{1.18}$$

$$m_1(z_1\dot{x}_1-x_1\dot{z}_1)+m_2(z_2\dot{x}_2-x_2\dot{z}_2)+m_3(z_3\dot{x}_3-x_3\dot{z}_3)=c_9 \tag{1.19}$$

By using these three integrals, we can reduce further the order of equations of motion from the $12th$ to $9th$ order.

1.4 The Integral of Energy

By taking the scalar products of equations (1.11), (1.12) and (1.13) by $\dot{\vec{r_1}}$, $\dot{\vec{r_2}}$ and $\dot{\vec{r_3}}$ respectively, and by adding we obtain the following equation.

$$\sum_{i=1}^{3}m_i\ddot{r}_i.\dot{v}_i=G\left\{\frac{m_1m_2}{r_{12}^2}\hat{r}_{12}.(\dot{\vec{r_1}}-\dot{\vec{r_2}})+\frac{m_1m_3}{r_{13}^2}\hat{r}_{13}.(\dot{\vec{r_1}}-\dot{\vec{r_3}})+\frac{m_2m_3}{r_{23}^2}\hat{r}_{23}.(\dot{\vec{r_2}}-\dot{\vec{r_3}})\right\} \tag{1.20}$$

where $\dot{\vec{r_1}} - \dot{\vec{r_2}} = -\dot{\vec{r}}_{12}$, $\dot{\vec{r_1}} - \dot{\vec{r_3}} = -\dot{\vec{r}}_{13}$ and $\dot{\vec{r_2}} - \dot{\vec{r_3}} = -\dot{\vec{r}}_{23}$

and $\hat{r}_{ij} = \frac{\vec{r_{ij}}}{r_{ij}}$, $\dot{\vec{r}}_{ij} = \frac{d}{dt}(\hat{r}_{ij} r_{ij}) = r_{ij}\dot{\hat{r}}_{ij} + \hat{r}_{ij}\dot{r}_{ij}$ for $i, j = 1, 2, 3; i \neq$j.

Since $\hat{r}_{ij}$ is a unit vector, $\dot{\hat{r}}_{ij}.\hat{r}_{ij} = 0$ and $\hat{r}_{ij}.\hat{r}_{ij} = 1$. By using these in equation (1.20), we obtain

$$\sum_{i=1}^{3} m_i \dot{r}_i.\dot{v}_i = -G\left\{\frac{m_1 m_2}{r_{12}^2}\dot{\vec{r}}_{12} + \frac{m_1 m_3}{r_{13}^2}\dot{\vec{r}}_{13} + \frac{m_2 m_3}{r_{23}^2}\dot{\vec{r}}_{23}\right\} \tag{1.21}$$

This can be written as

$$\frac{d}{dt}\left\{\frac{1}{2}\sum_{i=1}^{3} m_i \dot{\vec{r}_i}^2\right\} = G\frac{d}{dt}\left\{\frac{m_1 m_2}{r_{12}} + \frac{m_1 m_3}{r_{13}} + \frac{m_2 m_3}{r_{23}}\right\} \tag{1.22}$$

therefore the kinetic energy (K.E) of the system by definition is

$$T = \frac{1}{2}\sum_{i=1}^{3} m_i \dot{r_i}^2 = \frac{1}{2}\sum_{i=1}^{3} m_i v_i^2 \tag{1.23}$$

and we define the potential energy (P.E) of the system by

$$V = -G\left\{\frac{m_1 m_2}{r_{12}} + \frac{m_1 m_3}{r_{13}} + \frac{m_2 m_3}{r_{23}}\right\} \tag{1.24}$$

By integrating the equation (1.22), we obtain

$$\frac{1}{2}\sum_{i=1}^{3} m_i \dot{\vec{r}_i}^2 = G\left\{\frac{m_1 m_2}{r_{12}} + \frac{m_1 m_3}{r_{13}} + \frac{m_2 m_3}{r_{23}}\right\} + E \tag{1.25}$$

where E is a constant of integral.

By using equations (1.23) and (1.24) in equation (1.25), we get

$$T + V = E \tag{1.26}$$

This is the equation of energy integral.

Equation (1.26) represents the conservation of energy for the system of three masses. By using this integral, we can reduce further the order of equations of motion from the $9th$ to $8th$ order. Together with the elimination of the time and the ascending node, the order of equations of motion may be finally reduced to a system of the $6th$ order.

In general, one can not solve for the coordinates and velocity components of the three mass points as functions of the time given their values at some initial epoch. The same is true for the n-body problem. The general n-body problem is unsolvable in closed form for $n \geq 3$, i.e., given the coordinates and velocities of $n \geq 3$ mass-points moving under their own mutual gravitational attraction, one can not predict the motion for any arbitrary succeeding time interval. It can not be predicted in general whether two or more of the masses will collide in any arbitrary finite time interval, or whether some member of the system will escape from it. In the next section, we shall discuss certain special configurations of the masses in the three-body problem which do lead to the closed form of solutions.

1.5 Stationary Solutions of the Three-Body Problem

Lagrange obtained two particular solutions of the three-body problem in 1772. The three mass points m_1, m_2 and m_3 are initially assumed to be projected in the same plane. Then by a stationary solution, i.e., one in which the geometric configuration of three masses remains invariant with respect to the time. If the motion of the masses is such that their mutual distances from each other remain unchanged, the configuration simply rotates in its own plane around the center of mass. On the other hand, an expansion or contraction may take place which does not alter the shape of the pattern of points. We consider first the special case in which the three masses revolve in coplanar circular orbits around their center of mass with constant angular speed.

Let their position vectors relative to the center of mass C as shown in the Figure-1.2, be $\vec{r_1} = r_1\hat{r}_1$, $\vec{r_2} = r_2\hat{r}_2$, and $\vec{r_3} = r_3\hat{r}_3$. For circular motion r_1, r_2 and r_3 are constant. Here $\hat{r}_1$, $\hat{r}_2$ and $\hat{r}_3$ are unit vectors in the directions of $\vec{r_1}$, $\vec{r_2}$ and $\vec{r_3}$ respectively.

The total coplanar acceleration of any one of the mass points is given by (Smart, 1962)

$$\dot{v} = (\ddot{r} - r\dot{\theta}^2)u_r + 2(\dot{r}\dot{\theta} + r\ddot{\theta}u_\theta)$$

where u_r and u_θ are unit vectors in the radial and transverse directions, respectively. With the restrictions adopted here in the three-body problem, we have for each mass $\dot{r} = \ddot{r} = 0$, $\dot{\theta} = n$, the constant angular speed of revolution about C and $\ddot{\theta} = 0$.

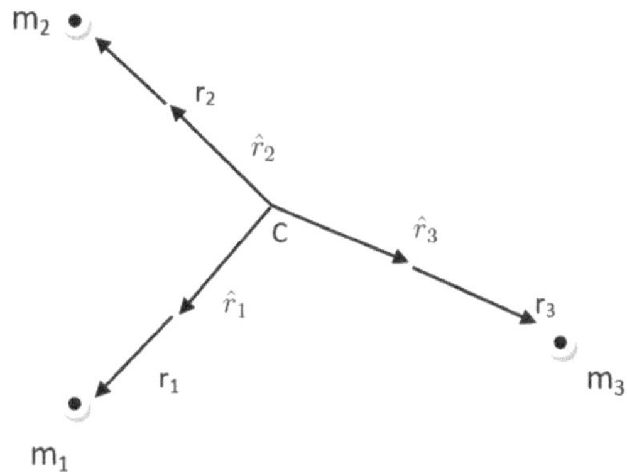

Figure 1.2: Orientation of the three - body system relative to center of mass

Therefore the accelerations become

$$\dot{v}_i = -r_i n^2 u_i, (i = 1, 2, 3)$$

Using these values in Equations (1.1), (1.2) and (1.3), we have

$$-n^2 r_1 u_1 = G\left\{\frac{m_2}{r_{12}^3}(r_2u_2 - r_1u_1) + \frac{m_3}{r_{13}^3}(r_3u_3 - r_1u_1)\right\} \tag{1.27}$$

$$-n^2 r_2 u_2 = G\left\{\frac{m_3}{r_{23}^3}(r_3u_3 - r_2u_2) - \frac{m_1}{r_{12}^3}(r_2u_2 - r_1u_1)\right\} \tag{1.28}$$

$$-n^2 r_3 u_3 = G\left\{-\frac{m_2}{r_{23}^3}(r_3u_3 - r_2u_2) - \frac{m_1}{r_{13}^3}(r_3u_3 - r_1u_1)\right\} \tag{1.29}$$

But the origin has been selected so that

$$m_1 r_1 u_1 + m_2 r_2 u_2 + m_3 r_3 u_3 = 0 \tag{1.30}$$

Therefore, we notice that by multiplying equations (1.27) by m_1, (1.28) by m_2 and adding, we obtain an equation which, upon substituting from equation (1.30), is equation (1.29). Hence we can use equation (1.30) in place of equation (1.28) as a third equation in the unit vectors u_1, u_2 and u_3.

We rewrite the system of equations in the form

$$\left. -n^2 + \frac{Gm_2}{r_{12}^3} + \frac{Gm_3}{r_{13}^3}\right) r_1u_1 - \frac{Gm_2}{r_{12}^3} r_2u_2 - \frac{Gm_3}{r_{13}^3} r_3u_3 = 0 \tag{1.31}$$

$$-\frac{Gm_1}{r_{12}^3} r_1u_1 + \left. -n^2 + \frac{Gm_3}{r_{23}^3} + \frac{Gm_1}{r_{12}^3}\right) r_2u_2 - \frac{Gm_3}{r_{23}^3} r_3u_3 = 0 \tag{1.32}$$

$$m_1 r_1 u_1 + m_2 r_2 u_2 + m_3 r_3 u_3 = 0 \tag{1.33}$$

These are the conditions which must be fulfilled if the three mass points are to move in a plane in circular orbits around the center of mass with uniform angular speed, n rad/sec.

Consider now a rectangular coordinate system with the origin at C (Figure-1.2) and rotating with angular speed n rad/sec in the counter clockwise direction. In this system, the unit vectors u_1, u_2, u_3 are fixed in position.

Let the angles in the equations $u_i = cos\theta_i i + sin\theta_i j$, $(i = 1, 2, 3)$ denote their orientations relative to these cartesian axes. Then by taking the scalar products of i and equations (1.31), (1.32) and (1.33) we have, after simplification,

$$-n^2 x_1 + \frac{Gm_2}{r_{12}^3}(x_1 - x_2) + \frac{Gm_3}{r_{13}^3}(x_1 - x_3) = 0 \tag{1.34}$$

$$-n^2 x_2 + \frac{Gm_1}{r_{12}^3}(x_2 - x_1) + \frac{Gm_3}{r_{23}^3}(x_2 - x_3) = 0 \tag{1.35}$$

$$m_1 x_1 + m_2 x_2 + m_3 x_3 = 0 \tag{1.36}$$

where $x_i = r_i cos\theta_i$, $(i = 1, 2, 3)$

similarly, using the scalar products of j and the same equations, we find

$$-n^2 y_1 + \frac{Gm_2}{r_{12}^3}(y_1 - y_2) + \frac{Gm_3}{r_{13}^3}(y_1 - y_3) = 0 \tag{1.37}$$

$$-n^2 y_2 + \frac{Gm_1}{r_{12}^3}(y_2 - y_1) + \frac{Gm_3}{r_{23}^3}(y_2 - y_3) = 0 \tag{1.38}$$

$$m_1y_1 + m_2y_2 + m_3y_3 = 0 \tag{1.39}$$

where $y_i = r_i sin\theta_i$, $(i = 1, 2, 3)$.

Equations (1.34) to (1.39) constitute a system of 6 simultaneous equations in the unknowns (x_i, y_i), (i=1, 2, 3).

We consider here the particular situation in which the masses are located at the vertices of an equilateral triangle. Then $r_{12} = r_{23} = r_{13} = r$ at all times.

Suppose we adjust the scale of length so that these are all unity for all values of the time. Equations (1.34) to (1.39) may be written then (with $G = 1$ by proper choice of time units)

$$(-n^2 + m_2 + m_3)x_1 - m_2x_2 - m_3x_3 = 0 \tag{1.40}$$

$$-m_1x_1 + (-n^2 + m_1 + m_3)x_2 - m_3x_3 = 0 \tag{1.41}$$

$$m_1x_1 + m_2x_2 + m_3x_3 = 0 \tag{1.42}$$

$$(-n^2 + m_2 + m_3)y_1 - m_2y_2 - m_3y_3 = 0 \tag{1.43}$$

$$-m_1y_1 + (-n^2 + m_1 + m_3)y_2 - m_3y_3 = 0 \tag{1.44}$$

$$m_1y_1 + m_2y_2 + m_3y_3 = 0 \tag{1.45}$$

These will have a non-trivial solution if

$$\begin{vmatrix} (-n^2+m_2+m_3) & -m_2 & -m_3 & 0 & 0 & 0 \\ -m_1 & (-n^2+m_1+m_3) & -m_3 & 0 & 0 & 0 \\ m_1 & m_2 & m_3 & 0 & 0 & 0 \\ 0 & 0 & 0 & (-n^2+m_2+m_3) & -m_2 & -m_3 \\ 0 & 0 & 0 & -m_1 & (-n^2+m_1+m_3) & -m_3 \\ 0 & 0 & 0 & m_1 & m_2 & m_3 \end{vmatrix} = 0$$

(obviously the trivial solution $x_i = 0$, $y_i = 0$ (i = 1, 2, 3) is inconsistent with the requirement here that $r = 1$).

Let $M = m_1 + m_2 + m_3$ denote the total mass of the system. Then by usual reduction rules for determinants, we find that the above determinant reduces to

$$m_3^2(M - n^2)^4 = 0 \tag{1.46}$$

This will be satisfied if $n^2 = M$. If we neglect G in the equations (1.34) to (1.39) and also the equal values of $r = r_{i,j}$ (i,j = 1, 2, 3; $i \neq$ j) then the above condition would be

$$n^2 = \frac{GM}{r^3} \tag{1.47}$$

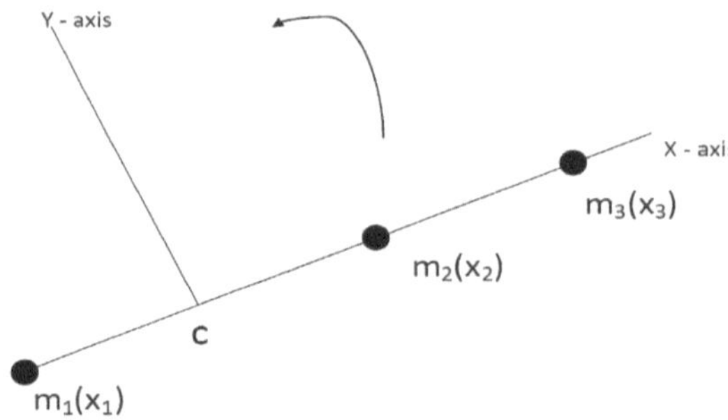

Figure 1.3: Straight - line solution in the three - body problem

Equation (1.47) is dimensionally consistent and indicates that n is in radians per unit time.

With $n^2 = M$ in equations from (1.40) to (1.45), we can choose any two pairs of coordinates (x_i, y_i) arbitrarily, adjust the scale, and determine the third pair in such a way that $r = 1$. The equations have a solution, therefore, in this special problem of three-bodies.

A second solution for equations (1.34) to (1.39) can also be found. Suppose that $y_1 = y_2 = y_3 = 0$. All three mass points lie on the x-axis. Then equations (1.37), (1.38) and (1.39) are satisfied. Let the masses be arranged on the x-axis as shown in the Figure - 1.3. We denote the distance $x_3 - x_2$ by r and arrange the scale so that $x_2 - x_1 = 1$. Then by virtue of the inequalities $x_1 < x_2 < x_3$, Equations (1.34), (1.35) and (1.36) become

$$n^2 x_1 + Gm_2 + Gm_3(1 + r)^{-2} = 0 \tag{1.48}$$

$$n^2(1 + x_1) - Gm_1 + Gm_3 r^{-2} = 0 \tag{1.49}$$

$$m_1 x_1 + m_2(1 + x_1) + m_3(1 + x_1 + r) = 0 \tag{1.50}$$

From these we eliminate n^2 and x_1 to obtain

$$(m_1+m_2)r^5+(3m_1+2m_2)r^4+(3m_1+m_2)r^3-(m_2+3m_3)r^2-(2m_2+3m_3)r-(m_2+m_3) = 0 \tag{1.51}$$

This is a fifth-degree equation in r, which has only one real positive root since there is only one change of sign in the coefficients. For the order in which the masses have been arrayed along the x-axis, this has only one positive solution for r. That is, when r is found, we may locate m_3 uniquely, because m_1 and m_2 have already been located arbitrarily to set the scale of distance $(x_2 - x_1)$. If we interchange the order of location of the masses, we obtain a similar unique solution for each arrangement. The solution for equation (1.51) can, in general, be found by numerical or iterative means.

Thus Lagrange has given two distinct solutions to the three-body problem, the equilateral triangle solution and the straight line solution, valid for any masses moving in coplanar circular orbits around their center of mass, with uniform angular speed. It is interesting to see the dynamical consequences of this type of motion, or, to see under what dynamical conditions such motions are likely to arise.

It is obvious that, because each mass moves uniformly in a circular path around the center of mass C, the areal velocity in each orbit must be constant. Hence the resultant force acting on each mass must pass through C. Equation (1.51) indicates that the acceleration to which each mass is subjected to $-r_i n^2 u_i$ (i = 1, 2, 3). Hence the force is $-m_i r_i n^2 u_i$. We see that the resultant force on each mass is directly proportional to the distance from C and is directed toward that point.

It is clear that in the discussion above that the initial velocity vectors required to produce circular motion of the system are perpendicular to the initial position vectors of the respective masses. Clearly, for each mass m_i the velocity $v_i = r_i \dot{u}_i$ so that the velocity is proportional to the distance of m_i from C and is perpendicular to u_i. The motion in the cases discussed above consists of a rotation of the entire system around the center of mass C.

If the initial velocity vectors are not perpendicular to the position vectors of their respective mass points, the configuration which initially may have existed will change with the time. If, however, each vector v_i (i = 1, 2, 3) makes the same angle with its corresponding r_i, then an expansion or contraction of the configuration will take place in such away that the problem is still solvable. The path followed by each mass will be a conic section of eccentricity depending on the initial circumstances of the motion. In this solution, the distances between the bodies vary with the time but always in such a way that their mutual distances apart remain in the same ratio. That is, if r_{12}, r_{23} and r_{13} denote the distances initially, at time t we would have αr_{12}, αr_{23} and αr_{13}, where α is the factor of proportionality. This means, that the distances of m_1, m_2 and m_3 from the center of mass at any time are $r_1 = \alpha r_{01}$, $r_2 = \alpha r_{02}$ and $r_3 = \alpha r_{03}$, where r_{0i}, $(i = 1, 2, 3)$ are the initial lengths of the position vectors. The shape of the configuration remains invariant with time. It is clear that α may be a function of the time. In this case that each mass traverses a conic section with a resultant pattern for the system as shown in Figure - 1.4 and 1.5 (Moulton, 1914).

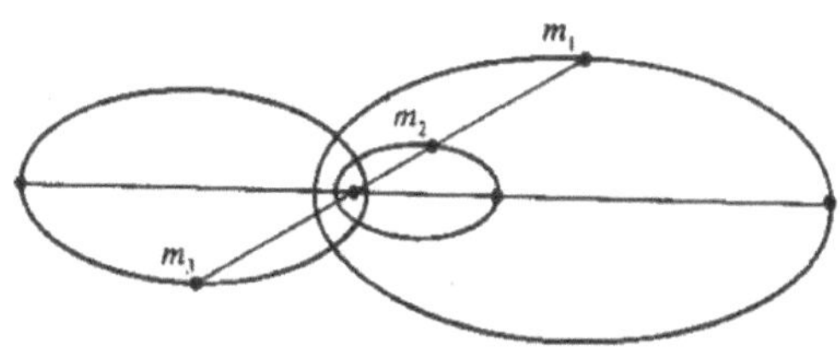

Figure 1.4: Straight - line solution in the three - body problem

The points at the vertices of the equilateral triangle solution of the three-body problem and the points of the straight line solution are called the Lagrangian points of the system. In a coordinate system rotating in its plane around the center of mass with constant angular speed n, these are fixed. A body situated at one of the Lagrangian points and initially at rest there will remain so unless disturbed by external forces. At these points the gravitational and centrifugal forces just balance each other.

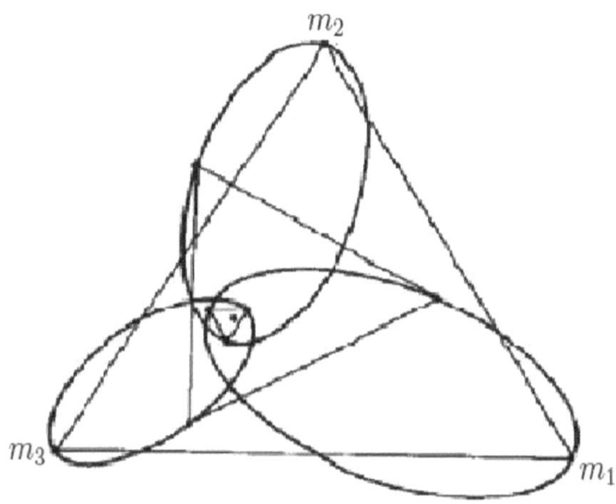

Figure 1.5: Equilateral - triangle solution in the three - body problem

Chapter 2

The Dynamics of the Sun-Earth-Moon

2.1 Introduction

The Sun-Earth-Moon system plays an important role in the application of the three-body problem for many reasons. It was the first three-body problem to be studied with Newtonian mechanics. It is now a very well-studied system. Hence it is a system that can be used as a standard system against which other triple systems may be compared. The dynamics of the Sun-Earth-Moon system is discussed with special attention to the effects of Sun's perturbations on the Moon's orbit around the Earth. Important secular effects are the regression of the nodes, the advance of the perigee and the increase in the Moon's mean longitude.

We discussed in this chapter about the relationship of the advance of perigee with the fluctuations in the Moon's synodic period. The study of the Moon has been particularly useful in the development of mathematics. The Earth and the Moon are sufficiently close to the Sun and hence the Moon's orbit around the Earth is perturbed by the gravitational attraction of the Sun. Smaller perturbations arise due to the attraction of other planets and also due to deviations of the shapes of the Earth and the Moon from perfect spherical symmetry.

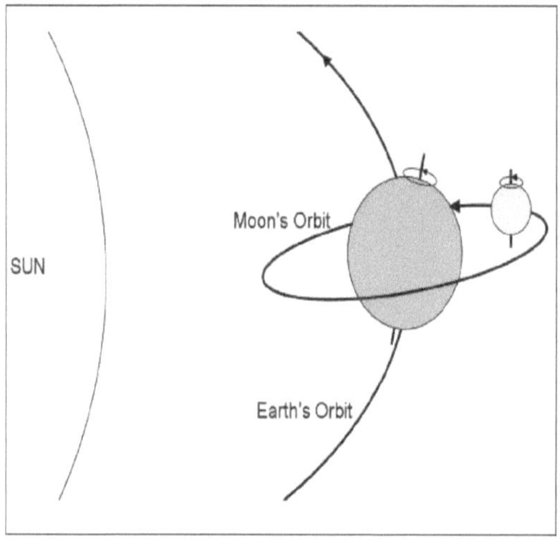

Figure 2.1: The Sun - Earth - Moon system

The Sun-Earth-Moon system (Figure - 2.1) has been a very important triple system of the celestial mechanics from the time of Newton, the discoverer of gravitation. Apart from Newton it has been studied by many great mathematicians like Lagrange, Laplace etc.,. Even in the simple three body case it poses lot of complexities. In the present study we try to understand the periodic fluctuations in the Moon's synodic period which may vary as much as one third of a day from the mean value. If the Earth and the Moon are considered as point masses and the effects of no other body is taken into account then all the six Keplerian orbital elements (shown in the Figure - 2.3) of the Moon's orbit i.e., semi major axis (a), eccentricity (e), inclination (i), longitude of the ascending node (Ω), argument of periapsis (ω) and mean motion (or period of revolution) (n) are invariant. But the actual situation is quite different. The Earth and the Moon are not point masses and the system is not an isolated one and hence the gravitational attraction of other bodies in the solar system influences their motion. As the Sun is most massive body and hence causes maximum perturbation on the orbital elements of the Moon compared to other bodies, we consider only the effect of the Sun. Because of departures from perfectly spherical shapes the gravitational forces cause the orbital elements undergo both periodic and secular changes (Danby, 1962; Moulton, 1970; Abhyankar, 1999).

2.2 Sun's Perturbations on the Moon's Orbit

Considering the Earth, the Moon and the Sun as a system of three mass points, the equation of motion of the Moon taking the Earth as the origin is given by (Danby, 1962):

$$\ddot{\vec{r}} + \frac{G(M_\oplus + m)\vec{r}}{r^3} = GM_\odot\left(\frac{\vec{\Delta}}{\Delta^3} - \frac{\vec{r'}}{r'^3}\right) \tag{2.1}$$

Where r is the radius vector of the Moon, r' is the radius vector of the Sun and $\Delta = r' - r$ is the radius vector of the Sun from the Moon (Figure - 2.2); $M_\oplus$, m and $M_\odot$ are the masses of the Earth, the Moon and the Sun respectively and G is the gravitational constant.

The second term on the left side in equation (2.1) is the principal force on the Moon due to the attraction of the Earth, while the term on the right side is the tidal force on the Moon due to

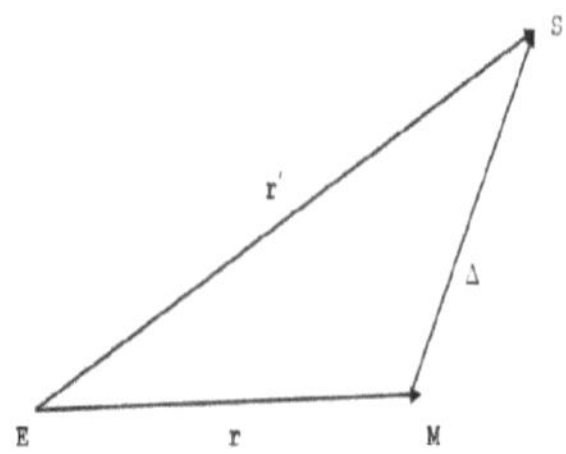

Figure 2.2: E - Earth, M - Moon, S - Sun three - body system

the Sun. The Sun is about $3, 30, 000$ times more massive than the Earth while its distance is only 400 times the distance of the Moon. With these values we find that the principal force is about 70 times the maximum tidal force.

The right hand side of equation (2.1) can be expressed as the gradient of the disturbing function R, given by:

$$R = GM_{\odot} \left(\frac{1}{\Delta} - \frac{\vec{r}.\vec{r'}}{\vec{r'}^3} \right) \tag{2.2}$$

Since $r/r' \approx 1/400$, we obtain an expression for R in terms of r/r' and then in terms of orbital elements. Some of the important terms are as follows:

$$\begin{aligned} R = \; & Gm\left(\frac{a_{moon}^2}{a_{sun}^3}\right)\left[\frac{1}{4} + \frac{3}{8}e^2 - \frac{1}{2}e\cos M - \frac{1}{8}e^2\cos 2M + \frac{15}{8}e^2\cos 2(\Omega + \omega - \psi') \right. \\ & - \frac{9}{4}e^2\cos\left[2(\Omega+\omega-\psi') + M\right] + \frac{3}{4}\cos\left[2(\Omega+\omega-\psi') + 2M\right] - \frac{15}{8}e^2\cos\left[2(\Omega+\omega-\psi') + 2M\right] \\ & \left. + \frac{3}{4}e\cos\left[2(\Omega+\omega-\psi') + 3M\right] + \frac{3}{4}e^2\cos\left[2(\Omega+\omega-\psi') + 4M\right]\right] \end{aligned} \tag{2.3}$$

Here a, e, i, Ω, ω are the Moon's orbital elements and M is its mean anomaly. The angle ψ' is the mean longitude of the Sun.

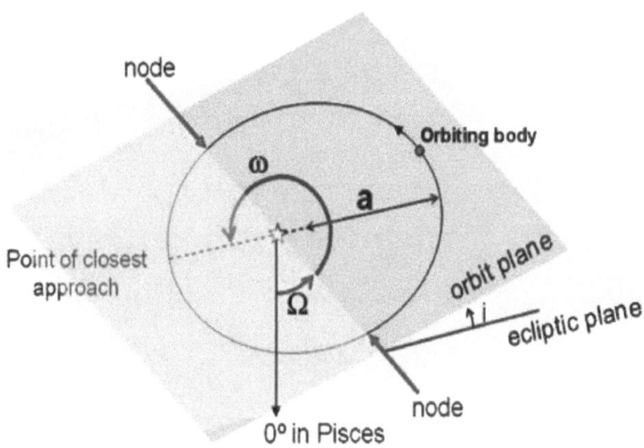

Figure 2.3: Orbital elements of the Moon

We can estimate the changes in the orbital elements with time by using Lagrange's planetary equations which relate the time derivatives of the elements with the derivatives of the disturbing function with respect to elements. We get:

$$\dot{a} = \frac{2}{na}\frac{\partial R}{\partial \sigma} \tag{2.4}$$

$$\dot{e} = \left.\frac{1-e^2}{na^2e}\right)\frac{\partial R}{\partial \sigma} - \left.\frac{\sqrt{1-e^2}}{na^2e}\right)\frac{\partial R}{\partial \omega} \tag{2.5}$$

$$\dot{\sigma} = -\left.\frac{1-e^2}{na^2e}\right)\frac{\partial R}{\partial e} - \frac{2}{na}\frac{\partial R}{\partial a} \tag{2.6}$$

$$\dot{\Omega} = \left.\frac{1}{na^2 \sin i\sqrt{1-e^2}}\right)\frac{\partial R}{\partial i} \tag{2.7}$$

$$\dot{\omega} = -\left.\frac{\cos i}{na^2 \sin i\sqrt{1-e^2}}\right)\frac{\partial R}{\partial i} + \left.\frac{\sqrt{1-e^2}}{na^2e}\right)\frac{\partial R}{\partial e} \tag{2.8}$$

$$\frac{di}{dt} = \left.\frac{\cos i}{na^2 \sin i\sqrt{1-e^2}}\right)\frac{\partial R}{\partial \omega} - \left.\frac{1}{na^2 \sin i\sqrt{1-e^2}}\right)\frac{\partial R}{\partial \Omega} \tag{2.9}$$

It may be convenient to use the mean anomaly $M = n(t-T) = nt + \sigma$ instead of σ itself, which is the coefficient of the secular acceleration. This implies that perturbation function R can be written in the form $R(a, e, i, \Omega,\ \omega,\ M)$.

The solution of these equations shows that the elements a, e, i, T undergo only periodic changes, while the elements $\Omega,\ \omega$ undergo both periodic and secular changes. The mean values of a, e and i are:

$a = 384400km,\ e = 0.05490$ and $i = 5^o09'$

The following important results of analytical studies may be noted (Danby, 1962; Abhyankar, 1999).

(a) a varies periodically by $\pm 0.9\%$, being maximum at the new moon and full moon and minimum at quadratures. Thus the period of perturbation is half a synodic month.

(b) The variation in e is about ± 0.0117 i.e. by about 20%. Eccentricity is the most seriously perturbed element in the periodic perturbations.

(c) The node moves backwards continuously and makes one complete revolution in 18.6 years. The variation from the average is $\pm 1^o 40'$. The regression of the node is related with the recurrence of eclipses of the Sun and the Moon.

(d) The perigee advances continuously and takes 8.85 years for a complete revolution. The variation from the average value is $\pm 12^o 20'$. The advance of perigee is related with the periodicity in the fluctuation of the Moon's Synodic period.

Thus the two important secular perturbations on the Moon's orbit due to the Sun are the advance of perigee and the regression of nodes. In this chapter, we focus on this aspect of the synodic period variation of the Moon and show how the changes in the angular separations of the longitudes of the perihelion of the Earth and the Moon at the time of the new moon are correlated.

2.3 Periodic fluctuations in moon's synodic period and advance of Moon's perigee

We have seen above that a changes periodically due to perturbations of the Sun and hence the period P of the Moon's orbit will also experience periodic changes. We know that the synodic period is the time taken by the Moon to reach the same phase again i.e., to say from one new moon to the next new moon. This period fluctuates between 29 and 30 days, the average being 29.530 588 86 days.

To study various characteristics of the fluctuations in the synodic period and its relationship with the advance of perigee, we have taken times of 630 new moons from January 2000 to December 2050 (Meeus, 1995). The difference between the actual synodic period (i.e., the time difference between two successive new moons) and its mean value, called the amplitude of fluctuation, ΔP which is plotted in Figure - 2.4.

From Figure - 2.4, it is clear that ΔP undergoes two sinusoidal nature of variations. There is a smaller cycle of about 14 synodic months and a larger cycle of about 112 (8×14) synodic months. The chosen interval of 630 synodic months covers five larger cycles. The following characteristics of variation in the period may be noted. The ΔP values starting from close to zero keep on increasing on the positive side and after attaining a maximum starts decreasing and reaches a peak on the negative side of $\Delta P$and then reversing the trend it reaches a value of ΔP

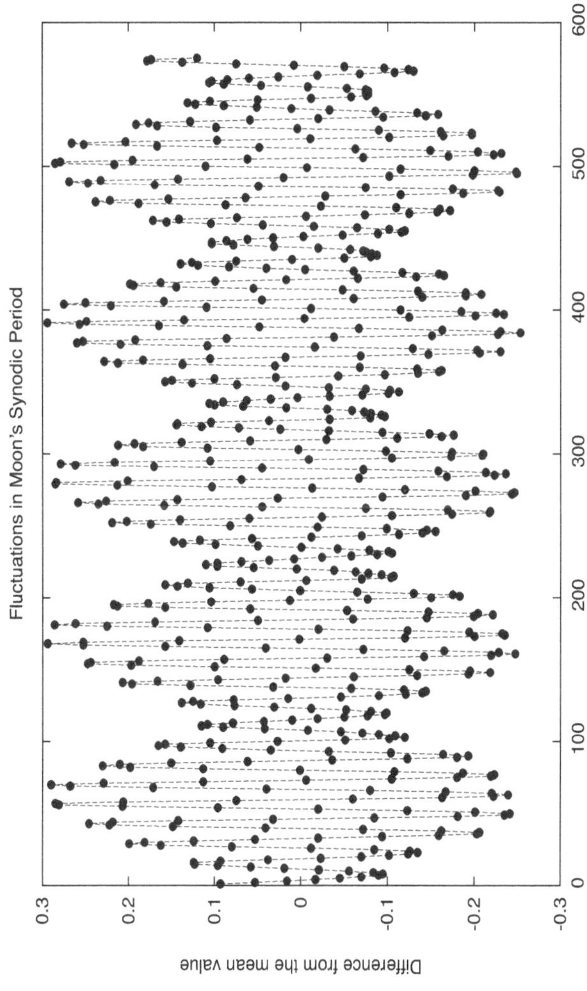

Figure 2.4: Moon's Synodic period from January 2000 to December 2050

close to zero again. This cycle is completed in about 14 synodic months. However the next cycle peak on either side of the mean attains a slightly higher maximum and minimum values. This continues for four consecutive cycles and then the trend reverses for the next 4 cycles. Thus eight such cycles form another gross fluctuation cycle (clearly seen by joining all the ΔP peaks of the smaller cycles) of about 112 synodic months. This period is close to the period of one complete revolution of Moon's perigee. We shall discuss these features qualitatively. The smaller cycle resembles a sine curve as shown in Figure - 2.5, while the larger cycle resembles a double sine curve as shown in Figure - 2.6.

In table 2.1 & 2.2, we given the value of ΔP for each of the eight peaks on either side of the mean for the five larger cycles along with various angular separations of interest at that time of the new moon. It may be observed that the peaks of a new larger cycle have slightly different amplitudes which increases and decreases with the increase and decrease in the value of the angular separation of the new moon and the Moon's perigee. This indicates a strong correlation between the two. It is clear that the larger fluctuation cycle depends upon the phase difference between the perigee of the Moon and perigee of the Sun, while the smaller cycle depends on the phase difference between the position of the new moon and the perigee of the Moon's orbit. In one synodic period the perigee of the Moon moves ahead by about 26^0. Hence it will require about 14 synodic months for the new moon to come the same position with respect to its perigee. This explains the smaller 14 synodic months cycle of fluctuation as shown in Figure - 2.5.

Table 2.1: Maximum deviations of the synodic period from the mean value

On the positive side of the mean | **On the negative side of the mean**

	Cycles	1/2/3/4/5						Cycles	1/2/3/4/5				
Date	ΔP	EMP	ESP	SMP	r	r'	Date	ΔP	EMP	ESP	SMP	r	r'
Peak No.1													
03/10/05	0.124	233	267	326	0.2652	1.005	08/02/05	-0.095	29	37	352	0.2445	0.9866
24/09/14	0.138	218	258	321	0.2678	1.0032	01/03/14	-0.099	41	58	344	0.2462	0.9908
14/10/23	0.147	230	278	312	0.2658	0.9975	20/02/23	-0.105	27	48	339	0.2443	0.9887
04/10/32	0.157	216	268	307	0.2683	1.0002	11/03/32	-0.113	39	68	330	0.2458	0.9935
25/10/41	0.171	227	288	299	0.2663	0.9945	01/04/41	-0.120	50	88	322	0.2477	0.9992
Peak No.2													
22/10/06	0.199	208	286	283	0.2693	0.9952	27/04/06	-0.135	57	115	303	0.2490	1.0067
11/11/15	0.207	220	306	274	0.2675	0.9900	18/05/15	-0.144	68	134	294	0.2514	1.0114
01/11/24	0.219	206	296	269	0.2696	0.9924	08/05/24	-0.155	54	125	289	0.2484	1.0093
22/11/33	0.228	218	317	261	0.2679	0.9877	28/05/33	-0.162	65	144	281	0.2507	1.0134
12/11/42	0.238	203	307	256	0.2699	0.9898	19/05/42	-0.172	51	135	276	0.2478	1.0117
Peak No.3													
09/12/07	0.246	211	334	237	0.2689	0.9849	15/06/07	-0.206	57	161	257	0.2490	1.0157
29/11/16	0.248	196	324	232	0.2705	0.9863	05/06/16	-0.218	43	152	251	0.2465	1.0146
20/12/25	0.259	208	345	224	0.2693	0.9838	25/06/25	-0.218	54	171	243	0.2484	1.0165
10/12/34	0.260	194	335	218	0.2706	0.9848	16/06/34	-0.230	40	162	238	0.2460	1.0158
31/12/43	0.269	206	356	210	0.2696	0.9833	07/06/43	-0.229	26	153	233	0.2442	1.0148
Peak No.4													
26/01/09	0.285	214	24	191	0.2685	0.9846	03/07/08	-0.241	32	178	214	0.2449	1.0167
17/01/18	0.294	199	14	186	0.2702	0.9837	24/06/17	-0.247	18	170	209	0.2435	1.0164
07/01/27	0.285	185	4	181	0.271	0.9833	14/07/26	-0.246	29	188	200	0.2445	1.0165
28/01/36	0.294	197	25	172	0.2704	0.9847	05/07/35	-0.253	15	180	195	0.2433	1.0167
18/01/45	0.285	182	15	167	0.2711	0.9838	25/06/44	-0.249	1	171	190	0.2429	1.0165

Table 2.2: Maximum deviations of the synodic period from the mean value (*cont..*)

On the positive side of the mean **On the negative side of the mean**

	Cycles	1/2/3/4/5						Cycles	1/2/3/4/5				
Date	ΔP	EMP	ESP	SMP	r	r'	Date	ΔP	EMP	ESP	SMP	r	r'
Peak No.5													
14/02/10	0.290	190	42	148	0.2708	0.9875	22/07/09	-0.239	7	196	171	0.2430	1.0160
04/02/19	0.286	176	33	143	0.271	0.9858	13/07/18	-0.235	353	187	166	0.2429	1.0166
25/02/28	0.279	188	53	153	0.2709	0.9898	02/08/27	-0.236	4	206	158	0.2429	1.0149
15/02/37	0.275	173	43	130	0.271	0.9877	23/07/36	-0.234	350	197	153	0.2430	1.0159
07/03/46	0.266	185	64	121	0.271	0.9924	12/08/45	-0.231	1	216	144	0.2429	1.0133
Peak No.6													
04/03/11	0.230	166	61	105	0.2706	0.9916	08/09/10	-0.223	7	242	125	0.2430	1.0075
24/03/20	0.217	178	81	97	0.2711	0.9971	30/08/19	-0.221	353	233	120	0.2429	1.0098
15/03/29	0.212	163	71	92	0.2704	0.9944	18/09/28	-0.210	4	253	112	0.2429	1.0046
04/04/38	0.198	175	91	84	0.2710	1.0001	09/09/37	-0.208	350	244	107	0.2430	1.0071
26/03/47	0.191	161	82	79	0.2702	0.9974	30/09/46	-0.197	1	263	98	0.2429	1.0071
Peak No.7													
21/04/12	0.165	168	108	59	0.2707	1.0050	27/09/11	-0.193	343	260	82	0.2434	1.0024
11/05/21	0.157	179	128	51	0.2711	1.0101	17/09/20	-0.183	329	251	77	0.2448	1.0050
02/05/30	0.144	164	119	46	0.2705	1.0078	07/10/29	-0.176	340	271	69	0.2437	0.9993
23/05/39	0.139	176	138	38	0.2710	1.0123	28/09/38	-0.165	326	262	64	0.2452	1.0020
11/06/48	0.131	187	158	29	0.2710	1.0154	19/10/47	-0.158	336	281	56	0.2439	0.9963
Peak No.8													
08/06/13	0.115	168	155	13	0.2707	1.0151	15/10/12	-0.120	318	279	39	0.2434	1.0024
29/06/22	0.110	179	174	5	0.2711	1.0166	04/11/21	-0.107	330	299	31	0.2446	0.9916
19/07/31	0.106	190	193	357	0.2708	1.0163	26/10/30	-0.097	316	290	26	0.2467	0.9940
08/08/40	0.103	201	212	348	0.2701	1.0140	16/11/39	-0.088	328	310	18	0.2449	0.9891
29/07/49	0.106	187	203	343	0.2710	1.0153	05/12/48	-0.078	340	310	10	0.2437	0.9854

Symbols used in the above table:

ΔP: Actual synodic period - value of mean synodic period.
EMP: Longitude of moon - Longitude of moon's perigee.
ESP: Longitude of sun - Longitude of sun's perigee.
SMP: Longitude of sun's perigee - Longitude of moon's perigee.
r: Distance from the Earth to the Moon in units of A.U.
r': Distance from the Earth to the Sun in units of A.U.

Note: Moon's distance is in Units of 1/100 A.U

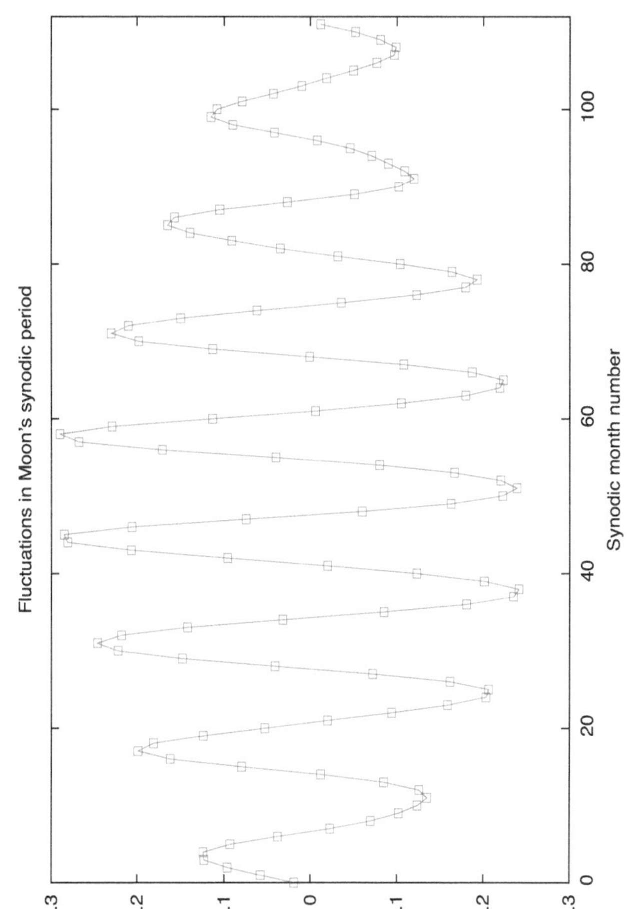

Figure 2.5: Enlarged view of first cycle out of five large cycles shown in Figure 2.4

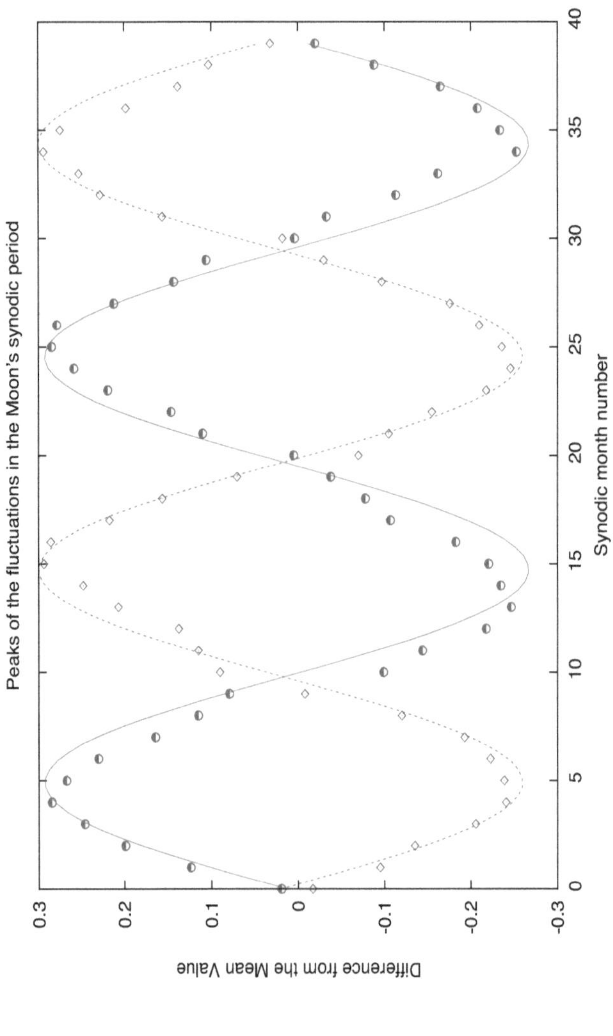

Figure 2.6: Double sine nature of Moon's synodic period of five large cycles shown in Figure 2.4

It is found that for the larger cycle the amplitude for each of the eight peaks depends on the angular separation of the new moon with the Moon's perigee, with the Sun's perigee and with the separation of Moon's and Sun's perigees.

2.4 Conclusions

We have seen above that 14 synodic months is the time interval for the new moon to recur at the same phase with its perigee. Further it may be noted that 8×14 synodic months ≈ 9 solar years. Therefore this is the time interval for the recurrence of both the lunar and solar perigees at new moon, which is the period of the larger cycle. Thus it is clear that there is a strong correlation between the angular separation of the new moon position with its perigee and the solar position. As these separation changes are periodic in nature we find the synodic period also shows periodic variations.

The distance of the Sun also shows some effect on the variation as it is found to increase on the positive side of ΔP up to maximum value then decreases, while on the negative side it increases for this part of the half cycle.

The phase difference, at the time of new moon, between perigee of the Moon & the Sun remains within 2°-4°, while the numerical value of the difference decreases during the entire cycle.

The difference between the longitudes of the Moon and that of it's perigee is about 26° at every new Moon in the entire cycle. While the difference between the longitudes of the Sun that of Sun's perigee lies between 28° to 29° at every new Moon in the cycle.

The changes in the semi-major axis due to perturbations of the Sun results in the changes in orbital period which in turn changes the energy of the orbit. Maximum energy occurs when the Sun is at perigee and the Moon is at apogee. Conversely the maximum decrease in energy with respect to the average occurs when the Sun is at apogee and the Moon is at perigee. The amplitude of various peaks on the positive and the negative sides of the mean can be understood by this underlying physics.

Chapter 3

Three-Body-Scattering Problem

3.1 Introduction

Generally the three-body systems are unstable. Frequently, they are only short lived stages in the evolution of a dynamical system. Typically, in a three-body encounter, two of them form a binary while the third escapes from the system. This interaction could be a distant flyby or a close encounter with one of the binary members. Both types of events are important.

In this chapter, we study the three-body scattering problem, first introduced by Burrau (1913), where three bodies start from rest at the vertices of a Pythagorean right-angled triangle. The bodies are assumed to be point masses and the Newtonian equations of motion are applied at close encounters. We perform simulation and investigate the nature of the solution for large intervals of time. We find that initially the three bodies interact forming double or triple systems which are short lived and ultimately the larger two bodies form a stable binary while the smallest body escapes the system.

The three-body scattering problem: The study of the motion of three mass points moving under mutual gravitational attractions is called the three-body problem. If one of the bodies is initially unbound or escapes to infinity after an interaction with the other two it is called the three-body scattering problem. The geometrical structure of Pythagorean problem of three-bodies is as shown in the Figure - 3.1.

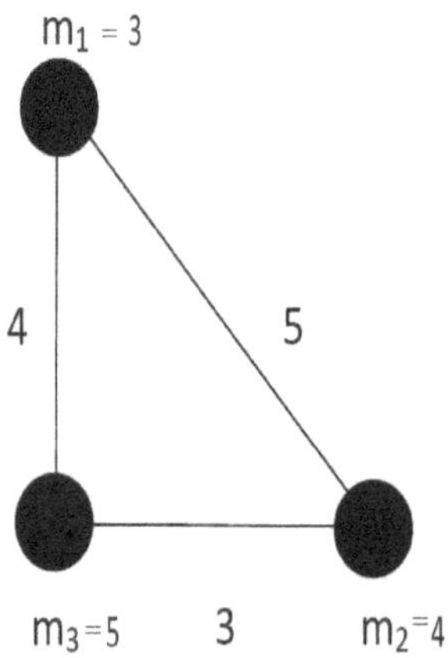

Figure 3.1: Pythagorean problem of three-bodies

3.2 Classification of the motion of the three bodies in a plane

Type1: Interplay:

In the first type of the three-body interaction which we call interplay, one of the participating bodies performs repeated and close approaches with the other two bodies. In most cases, the largest mass is the central body which performs close approaches with the two other bodies. The duration of the existence of this configuration may be arbitrary sooner or later a new configuration evolves.

Type2: Ejection without Escape:

In this configuration a close approach occurs such that one of the participating bodies receives sufficient energy to depart from the other two bodies with negative relative energy. Therefore, it does not escape but it departs on an highly elliptic orbit. The remaining two bodies form a binary. The time required for the ejected body to return depends on its relative energy and it can vary greatly. Upon its return to the binary, the ejected body might start to participate in an interplay motion as in type1, it might be again ejected immediately, or finally, it might form a binary with one of the bodies which were left behind during its previous ejection. Then another member of the three bodies might be ejected. In this type of motion, therefore, it might follow an interplay or it might end in an interplay. If the ejected body receives sufficient energy for escape we face the third possible configuration.

Type3: Escape:

By this means that the ejected body has sufficient energy to assume a hyperbolic orbit with respect to the binary. This is a possible evolution from interplay motion and it is a positive energy version called as Ejection.

Type4: Revolution:

The binary is formed by two members of the three bodies and the orbit of the third body surrounds the orbit of the binary. The orbit of the third body might be circular or elliptic and so might be the orbits of the participating members of the binary. If the ratio of the semi-major axis of the orbit of the third body to the semi-major axis of the binary is sufficiently large, a stable revolution of the system may be expected. If this ratio is below a certain limit, which depends on the masses and on the orbits of the participating bodies as well as on the phase between the motion, the revolution becomes an interplay.

Type5: Resonance:

If the total energy of the three-body system is negative, a temporary capture may occur, leading to prolonged interplay of the three bodies. This event is called resonance.

Type6: Ionization:

All three bodies become unbound and escape to infinity in different directions with carry off kinetic energy.

3.3 Study of motion of three-bodies with numerical results

Burrau's problem of three bodies is described in the following way. Three mass points attract each other according to the Newtonian law of gravitation. The three mass points m_1, m_2 and m_3 are placed on the vertices of a right angled triangle of sides in ratio 3:4:5 and the three masses are also in the same ratio. They are placed such that $m_1 = 3$ is opposite side of length 3, $m_2 = 4$ is opposite side of length 4 and $m_3 = 5$ is opposite side of length 5 as shown in Figure - 3.1.

The particles are free to move in the plane of the triangle and are at rest initially. The problem treated in this chapter is to determine their subsequent motion. Because of the relation of the initial conditions to a Pythagorean triangle, Burrau's problem may also referred to as a Pythagorean problem of three bodies.

In this chapter, the historical and scientific background, the concept of its solution, and the numerical results are discussed. Finally, we concluded that a stable binary formed after a long duration of an interaction of three bodies which means that the two larger bodies form a binary and the smallest body escapes from the system.

Meissel (1893) suggested one of the special cases of three body problem, the same was analyzed by Burrau (1913). The problem has been modified several times as suggested by Standish (1966). The formation of a close binary, the escape of one of the masses, several close approaches and other qualitative aspects of the solution have applications to dynamical evolution of star cluster.

3.4 Historical and scientific background

Historically, various attempts have been made to obtain periodic, quasi-periodic and asymptotic solutions of the three bodies under proper suited initial conditions or simplifying by imposing restrictions and assumptions. The different simplifying assumptions lead to the well-known lunar and planetary theories from Newton to Eckert, Clemence and Brouwer (1961). The works of Arnold (1963), Moser (1966), Jefferys and Moser (1966) and Arenstorf (1966) given periodic and quasi-periodic solutions.

Assuming that the participating three bodies have almost equal masses and using the same configuration as described above a totally separate type of dynamical system. We have to search the initial conditions that leads to the periodic, quasi-periodic or asymptotic solutions for this group of problems.

The classical examples regarding periodic motions were discovered by Lagrange (1772). The generalizations of these solutions were given by Wintner (1941), Whittaker (1900) and Lovett (1911). Stromgren's (1900) independent contributions from 1900 to 1922 should be mentioned, not only because of their importance but also because of their unquestionable influence on Burrau's problem. The insight into finding solutions to such problems has not changed significantly since 1922 and we may conclude that the existence of periodic orbits in the general problem of three bodies, when no restrictions are placed on the masses and distances, seems to be still limited essentially to orbits of librational type. Stromgren (1922) worked on this problem and given an asymptotic orbit related solution. This type of motion is known as an elliptic hyperbolic orbit (Merman,1958).

Burrau (1913) was interested in finding such periodic solutions for the general problem of three bodies, without restrictions on the masses and on the distances, such solutions were not of librational type. As mentioned by Burrau (1913), Meissel (1893) was of the opinion that the initial conditions described by the Pythagorean problem of three bodies will result in a periodic motion. Further, Burrau (1913) performed a numerical integration with variable step size for a limited time and found no special property of the motion.

3.5 Concept of solution

The motion of the particles is restricted to the plane of the triangle is an immediate outcome of the initial conditions. Since the angular momentum of the system L at the beginning of the motion is zero, it will continue in the same state during the motion. The necessary but not sufficient condition for the simultaneous collision of all three particles is that angular momentum is zero i.e., $L = 0$. Proof of the theorem has been given at end of the chapter. Therefore, the theorem does not predict triple collisions but permits its possibility. The other integrals of motion namely the center of mass and of the energy, are applicable without any complication, but they do not contribute to obtain a solution. One practical approach to the existence of this problem is numerical simulations. Introducing variables which regularize binary collisions, numerical integrations are performed at several levels of accuracy and the results of these integrations are analyzed for the occurrence of triple collisions. If there appear only binary collisions, the solution is considered established.

The existence of the solution is established once the occurrence of triple collisions is excluded. This is certainly the case when $L \neq 0$ because of the previously mentioned necessary condition for triple collisions. In our problem $L = 0$, therefore the existence of the solution must be first established along the experimental lines indicated previously. Assuming that the occurrence of triple collisions may be excluded, reference is made to Poincare (1884) and Sundman (1912) to their representation of the solution in power series, convergent for all values of the time. Such definition or existence of the solution is of questionable for the description of the behavior of a dynamical system. In general, from such series representations, we can not understand the characteristic features of the motion and its details qualitatively or quantitatively.

Solutions obtained from high-speed electronic computers have disadvantages similar to those of representations by series, since the description of the behavior of the system over a long period of time may not be expected realistically because of the accumulation of end-figure errors.

Consequently, the behavior of a dynamical system may be described and understood only in certain special cases. When the initial conditions are such as to lead to periodic, quasi-periodic or asymptotic motion, then the qualitative and often also the quantitative aspects of the solution are settled. This is one reason for the search for special sets of initial conditions that result in the above mentioned types of motions. For the purpose of this problem, a solution is defined as follows. The solution is a qualitative and quantitative description of the general and detailed behavior of a dynamical system for $t = -\infty$ to $+\infty$. Birkhoff (1920) given importance to the qualitative aspects and the general behavior rather than the numerical results since the main purpose of dynamics is to understand the over all performance and activity of a system.

3.6 Solution

Let a system of n-bodies consist of point masses m_i at $\vec{r_i}$ where $i = 1, 2, ..., n$ and the $\vec{r_i}$ are expressed with respect to an inertial frame of reference.

Let $r_{ij} = |\vec{r_i} - \vec{r_j}|$, $(r_{ij} = r_{ji})$, then the equation of motion of the particle m_i is

$$m_i \ddot{\vec{r}}_{ij} = -Gm_i \sum_{i=1}^{n} \frac{m_i(\vec{r_i} - \vec{r_j})}{r_{ij}^3} \tag{3.1}$$

and $j \neq$i. Now we consider the general expressions for the motion of three mass points m_1, m_2 and m_3 are (scale is so chosen such that G = 1).

$$\ddot{\vec{r_1}} = -m_2 \frac{\vec{r_1} - \vec{r_2}}{r_{12}^3} - m_3 \frac{\vec{r_1} - \vec{r_3}}{r_{13}^3} \tag{3.2}$$

$$\ddot{\vec{r_2}} = -m_3 \frac{\vec{r_2} - \vec{r_3}}{r_{23}^3} - m_1 \frac{\vec{r_2} - \vec{r_1}}{r_{21}^3} \tag{3.3}$$

$$\ddot{\vec{r_3}} = -m_1 \frac{\vec{r_3} - \vec{r_1}}{r_{31}^3} - m_2 \frac{\vec{r_3} - \vec{r_1}}{r_{32}^3} \tag{3.4}$$

where,

$$r_{12} = \sqrt{(x_1 - x_2)^2 + (y_1 - y_2)^2 + (z_1 - z_2)^2} \tag{3.5}$$

$$r_{13} = \sqrt{(x_1 - x_3)^2 + (y_1 - y_3)^2 + (z_1 - z_3)^2} \tag{3.6}$$

$$r_{23} = \sqrt{(x_3 - x_2)^2 + (y_3 - y_2)^2 + (z_3 - z_2)^2} \tag{3.7}$$

Since the general problem of three bodies will be described by second order vector differential equation for each particle, there will be 18 constants of motion. The conservation of angular momentum and the energy together with the uniform motion of center of mass will provide 10 constants leaving eight to be determined. Since the general potential affecting anymore of the objects will not be that of a single point mass we should not expect the orbits of the objects to close and we are left with eight arbitrary constants required to specify the problem. To be sure the initial position and velocities of the components would provide the 18 constants required for the unique solution of the motion, since the laws of Newtonian mechanics are deterministic. But these initial values are not integrals of the motion. The parameters they specify are not constant during the motion of the members of the system. Thus they provide a basis for calculating the motion of a specific system, they do not allow for a general solution.

Initially we set positions of three bodies accordingly Burrau (1913) and the initial velocities are set to be zero. Due to mutual gravitational attraction force, the velocity, acceleration and position of three bodies will change at each step and we have to substitute these values to get new position of the three particles. The following equations will give the new position of body at each iteration.

$$x_{1_n} = x_{1_i} + v_{x_{1_i}} \times dt \tag{3.8}$$

$$y_{1_n} = y_{1_i} + v_{y_{1_i}} \times dt \tag{3.9}$$

$$z_{1_n} = z_{1_i} + v_{z_{1_i}} \times dt \tag{3.10}$$

Similarly, we can write the expressions for the other coordinates of the mass points m_2 and m_3 such as x_{2_n}, y_{2_n}, z_{2_n} and x_{3_n}, y_{3_n}, z_{3_n}. Where the new velocities of each component can be calculated as follows.

$$v_{x_{1_n}} = v_{x_{1_i}} + a_{x_{1_i}} \times dt \tag{3.11}$$

$$v_{y_{1_n}} = v_{y_{1_i}} + a_{y_{1_i}} \times dt \tag{3.12}$$

$$v_{z_{1_n}} = v_{z_{1_i}} + a_{z_{1_i}} \times dt \tag{3.13}$$

Similarly, we can find the new velocities of other components of m_2 and m_3 such as $v_{x_{2_n}}$, $v_{y_{2_n}}$, $v_{z_{2_n}}$ and $v_{x_{3_n}}$, $v_{y_{3_n}}$, $v_{z_{3_n}}$. And also we can find the acceleration of each component by using equations of motion (3.2), (3.3) and (3.4).

Continue this procedure until to get the required accuracy of system. The following numerical result gives the approximate configuration of the three body-system. The complete solution from $t = -\infty$ to $+\infty$ may be described as follows. After few close interactions between three bodies, the two bodies m_2 and m_3 form a close binary. The center of mass of these two bodies leaves in a fixed direction from the center of mass of the three-body system, while m_1 leaves in the opposite direction. At $t = 0$, the velocities of participating bodies are zero, therefore the motion is completely symmetric with respect to this particular period of time. Which means that if $x(t)$ shows the position of the system at time t, then $x(t) = x(-t)$. Therefore, if we find the motion from $t = 0$ to ∞, it can also extend from $t = -\infty$ to $+\infty$. The hyperbolic departure of the close binary and of the mass m_1 in opposite directions as $t \to \infty$ completely describes the solution. The influence of m_1 on the binary is decreasing to zero as $t \to \infty$; due to this, the elliptic orbits of m_2 and m_3 are attained asymptotically.

By observing above, we describe the nature of the three-body system from $t = -\infty$ to $+\infty$ in the following way. At $t = -\infty$, the two bodies m_2 and m_3 form a close binary and the center of mass moves with hyperbolic velocity toward the body m_1. Due to interaction of m_1, the binary formed by m_2 and m_3 is dissolved at $t = -t_0 < 0$, and at time $t = 0$ the three bodies located at initial position described Pythagorean initial conditions with zero velocities. At $t = t_0$ the binary is formed by two bodies m_2 and m_3 and the hyperbolic separation of the body m_1 from the binary begins to take place and continues as $t = +\infty$. Between the time $t = -t_0$ and $t = t_0$ an interesting interaction occur between the largest body, $m_3 = 5$ and the two smaller bodies. Many times the largest body and the two smaller bodies to come very close, sometimes rotates alternating also, however no close approach occurs between the two smaller bodies.

One of the close approaches between m_3 and m_2 is approximately a collision with nearly rectilinear orbits. At that time the velocity of the smallest body m_1 is nearly zero. This combination of incidents suggests the existence of a set of initial conditions leading to a periodic solution. This solution might start with zero velocities of the participating bodies - not in the Pythagorean configuration. At $t = T/2 < t_0$ when m_2 and m_3 collide and the velocity of m_1 is zero, the solution reaches its half period. Then the motion is completely reversed and the bodies return to their original positions with zero velocity at $t = T$.

From the point of view of stellar dynamics, this example demonstrates capture, binary formation, resolution of a binary, escape, close approaches, possible collisions and interaction of a binary with a single star.

3.7 Conclusion

We performed simulation to investigate the nature of the solution of three-bodies for large intervals of time as shown in Figures - 3.2 and 3.3. We observed that initially the three bodies interact forming double or triple systems which are short lived and ultimately the larger two bodies form a stable binary while the smallest body escapes the system.

We have performed the simulation with the same code for other triple system such as equal masses placed on vertices of equilateral triangle and observed that the three bodies collide at centriod of triangle. Initially triple system is at rest, as mentioned necessary condition for simultaneous triple

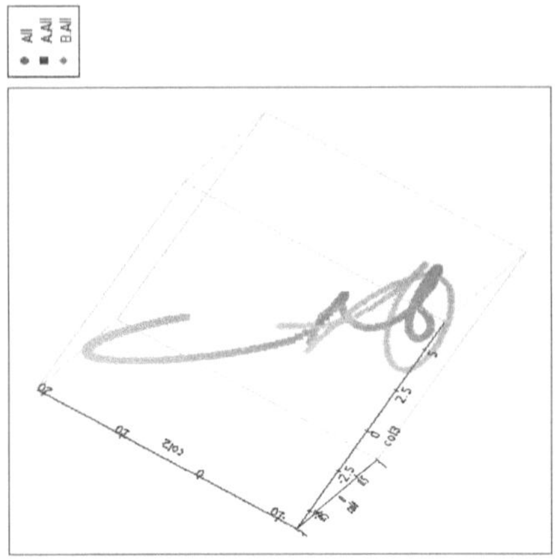

Figure 3.2: Interaction of three bodies m_1 = 3 (blue), m_2 = 4 (red) and m_3 = 5 (green) for small time interval

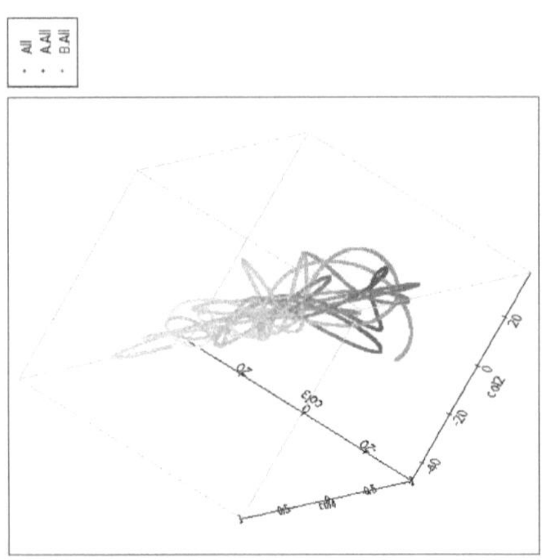

Figure 3.3: Interaction of three bodies m_1 = 3 (blue), m_2 = 4 (red) and m_3 = 5 (green) for large time interval

collision is that the total angular momentum of system is zero. Our numerical simulation illustrates the result. In general a three - body system is unstable. In a triple system most of the time the smallest mass escapes the system while the larger two form a stable binary. This is observed in star clusters in which a large number of binaries are formed as an end product of triple star encounters.

Theorem:

The necessary but not sufficient condition for the simultaneous collision of all three particles is that angular momentum is zero i.e., L = 0.

Proof: Let the components of angular momentum be

$$\alpha = \sum m_k(y_k \dot{z}_k - z_k \dot{y}_k)$$

$$\beta = \sum m_k(z_k \dot{x}_k - x_k \dot{z}_k)$$

$$\gamma = \sum m_k(x_k \dot{y}_k - y_k \dot{x}_k)$$

if at $t = t_1$, all n bodies collide at one point, then the three angular momentum constants α, β and γ must be zero. Since the center of mass rest at the origin, a collision of all n bodies can occur only at origin O.

Using the equation of moment of intertia,

$$I = \sum m_k r_k^2 = \sum m_k (x_k^2 + y_k^2 + z_k^2)$$

For $\tau \leq t < t_1$, we have

$$\frac{1}{2}\dot{I} = \sum m_k r_k \dot{r_k} \tag{3.14}$$

$$\frac{1}{2}\ddot{I} = \sum m_k (\dot{r_k}^2 + r_k \ddot{r_k})$$

$$= 2T + \sum r_k U_{r_k}$$

This leads to the Lagrange's formula (also known as the Virial theorem)

$$\frac{1}{2}\ddot{I} = 2T - U$$

Which can be expressed with the help of the energy integral $T - U = E$, where E denotes the total energy.

$$\frac{1}{2}\ddot{I} = T + E = U + 2E \tag{3.15}$$

At first we do not assume that all particles collide at $t = t_1$, where t_1 is a singularity for at least one coordinate $r(t)$. As we have seen, $U \to \infty$ as $t \to t_1$, and therefore, if $t_2 < t_1$ is sufficiently close to t_1, the right hand side of equation (3.15) will be positive throughout the interval $t_2 \leq t < t_1$. Hence $\ddot{I} > 0$ $(t_2 \leq t < t_1)$, and the function $\dot{I}$ is monotonically increasing. We may also assume that $\dot{I}$ is always positive or always negative in this interval, to change the sign at $t = t_3$, we need to replace t_2 by a number between t_3 and t_1. Thus the positive function I is monotonic in the interval $t_2 \leq t < t_1$, and therefore has a limit $t \to t_1$. From the definition of I this limit is zero if and only if all n bodies collide at the origin at $t = t_1$.

Using the algebraic identity

$$\sum \xi_k^2 \sum \eta_k^2 = (\sum \xi_k \eta_k)^2 + \sum (\xi_k \eta_1 - \xi_1 \eta_k)^2$$

$k = 1$ to $3n$ and $k < 1$

with $\xi_k = r_k \sqrt{m}$, $\eta_k = \dot{r_k} \sqrt{m}$, we obtain from the equation (3.14),

$$2IT = \frac{1}{4} \dot{I}^2 + \sum (\xi_k \eta_1 - \xi_1 \eta_k)^2 \tag{3.16}$$

expanding the equation (3.16) and retaining the last sum only those terms in which ξ_k and η_1 refer to the same particle, then we obtain,

$$2IT \geq \frac{1}{4} \dot{I}^2 + \sum m_k^2 \{(y_k \dot{z}_k - z_k \dot{y}_k)^2 + (z_k \dot{x}_k - x_k \dot{z}_k)^2 + (x_k \dot{y}_k - y_k \dot{x}_k)^2\}$$

on the other hand, α, β, γ in conjunction with the Schwarz inequality $(\sum x_i y_i)^2 \leq (\sum x_i^2)(\sum y_i^2)$ implies that

$$\alpha^2 = \{\sum m_k(y_k\dot{z}_k - z_k\dot{y}_k)\}^2 \leq n\sum m_k^2(y_k\dot{z}_k - z_k\dot{y}_k)^2$$

with analogous inequalities for β, γ so that the previous estimate gives

$$2IT \geq \frac{1}{4}\dot{I}^2 + \eta, \qquad \eta = \frac{\alpha^2 + \beta^2 + \gamma^2}{n} \tag{3.17}$$

for the proof of theorem, instead of equation (3.16), it is sufficient to consider the weaker inequality $2IT \geq \eta$ which by (3.14) is equivalent to

$$\ddot{I} \geq \frac{\eta}{I} + 2E \tag{3.18}$$

If I is monotonically decreasing in the interval $t_2 \leq t < t_1$, multiplying the equation (3.17) by the positive quantity $-2\dot{I}$ and integrating from t_2 to t, we obtain

$$\dot{I}_2^2 - \dot{I}^2 \geq \eta log\frac{I_2}{I} + 4E(I_2 - I)$$

where I_2, $\dot{I}_2$ denote the values of I, $\dot{I}$ at $t = t_2$

therefore certainly

$$2\eta log\frac{I_2}{I} \leq \dot{I}_2^2 + 4|E|I_2, \qquad (t_2 \leq t < t_1) \tag{3.19}$$

and when $\eta > 0$, i.e., α, β, γ are non zero simultaneously, this results in a positive lower bound for I. On the other hand, I is monotonically increasing for $t_2 \leq t < t_1$, then trivially we have $I \geq I_2$ in this interval.

Thus in both cases, if $\eta > 0$, the function I has a positive lower bound in the interval $t_2 \leq t < t_1$. Denoting the distance from the center of mass, the origin of co-ordinate system, to the point P_k by R_k, we have

$$I = \sum m_k R_k^2$$

more over

$$\sum m_k r_k = 0$$

and therefore

$$\sum m_k(r_l - r_k) = M r_l, \qquad M = \sum m_k$$

$$Mrl^2 \leq \sum m_k(r_l - r_k)^2 \qquad (l = 1, 2, ..., n)$$

$$MI \leq 2\sum m_k m_l r_{kl}^2, \qquad l < k$$

Consequently, if $\eta > 0$, the maximum of the $\frac{n(n-1)}{2}$ distances r_{kl} $(k < l)$ has a positive lower bound for $t_2 \leq t < t_1$, and also for $\tau \leq t < t_1$, and it follows that collision of all n bodies at O can not occur at $t = t_1$. This proves the theorem.

Chapter 4

Scattering phenomenon in binary systems

4.1 Introduction

Three-body-scattering plays an essential role in stellar dynamics. In particular the interactions between binary stars and single stars in globular clusters are thought to be an important dynamical process. The important changes in orbital elements are i.e., angular momentum, longitude of ascending node of the binary and the energy exchange, the eccentricity of the binary caused by an encounter with another star. In some cases even very small changes in orbital elements i.e., eccentricity or angular momentum can be very significant. The effect of stellar encounter on a binary star has been a subject of extensive study (Hut,1984; Heggie,1988 and Valtonen,1988).

Heggie and Rasio (1996) studied the effect of stellar encounters on the eccentricity of binaries. They discussed the change in eccentricity analytically for the cases, dependence on the periastron distance, angular and phase dependences, masses etc. Huang and Valtonen (1987) obtained an analytical expression for the energy transfer in a parabolic encounter of a single star with a circular binary keeping terms of the first order in a/q, where a is the radius of the circular orbit and q is the distance of closest approach of the single star with respect to the centre of mass of the binary. Numerically, they also obtained the energy transfer and change in eccentricity for the case of mass of the binary equal to the mass of the single star.

In this chapter, we deal with two cases, in first case we extended the analytic treatment of Huang and Valtonen (1987) to obtain angular momentum transfer and change in eccentricity for the case of a parabolic encounter of a star with a circular binary. In the second case, we derived analytical expression for the change in orbital eccentricity of a binary following a distant encounter with a third star on a hyperbolic or parabolic orbit. To verify the accuracy of the analytical expressions of Heggie and Rasio (1996), we given the detailed numerical results.

4.2 The effect of parabolic encounter on the angular momentum and eccentricity in a circular binary

4.2.1 Perturbation theory

The masses m_1 and m_2 move in a circular orbit about their common centre of mass. A third body of mass m_3 moves in a parabolic orbit inclined at an angle i to the orbital plane of the binary. The node line is used as the direction of reference. The angle subtended by the line joining m_1 and m_2 with the node line is denoted by Ω and the pericentre direction of m_3 is denoted by ω.

We consider time t from the instant m_3 is at the pericentre. At time $t = 0$, om_1 is taken as the ox-axis. Let $\mathbf{r}$ represent the position vector of m_2 relative to m_1 and $\mathbf{R}$ be the position vector m_3 relative to the center of mass binary m_2 and m_1 and let r and R be their moduli. Let E and e be the energy and the eccentricity of the relative orbit of the binary. The energy E will satisfy the following equation.

$$\dot{E} = (\dot{\mathbf{r}}, \ddot{\mathbf{r}}_p) \tag{4.1}$$

where $(\mathbf{a}, \mathbf{b})$ is the scalar product of vectors $\mathbf{a}$ and $\mathbf{b}$, and $\ddot{\mathbf{r}}_p$ is the perturbational acceleration from the third body.

The eccentricity e is connected with the Laplace vector ε which is defined by Roy (1978)

$$\varepsilon = \dot{\mathbf{r}} \times \mathbf{h} - \mu_{12}\frac{\mathbf{r}}{r} = \mu_{12} e \mathbf{p}$$

where $\mathbf{p}$ is a unit vector along the direction of the pericentre, $\mu_{12} = G(m_1 + m_2)$ and G is the gravitational constant, and $\mathbf{h} = \mathbf{r} \times \dot{\mathbf{r}}$ is the angular momentum per unit mass. Taking the derivative with respect to time, we have

$$\dot{\mathbf{h}} = \mathbf{r} \times \ddot{\mathbf{r}}_p$$

$$\dot{\varepsilon} = 2\mathbf{r}(\dot{\mathbf{r}}, \ddot{\mathbf{r}}_p) - \dot{\mathbf{r}}(\mathbf{r}, \ddot{\mathbf{r}}_p) - \ddot{\mathbf{r}}_p(\mathbf{r}, \dot{\mathbf{r}})$$

Then we can derive the equation of e by use of the relation, $\mu_{12}^2 e\dot{e} = (\varepsilon, \dot{\varepsilon})$. Since ε has a factor of e, the last term in ε can be neglected when calculating $(\varepsilon, \dot{\varepsilon})$ for a near-circular orbit. Finally, we have

$$\dot{e} = \frac{2}{\mu_{12}}(\mathbf{r}, \mathbf{p})(\dot{\mathbf{r}}, \ddot{\mathbf{r}}_p) + \frac{1}{\mu_{12} r e}(\mathbf{r}, \dot{\mathbf{r}})(\mathbf{r}, \ddot{\mathbf{r}}_p). \tag{4.2}$$

If we consider only the terms with the lowest power of (r/R) in the disturbing function, the expression of $\ddot{\mathbf{r}}_p$ is

$$\ddot{\mathbf{r}}_p \simeq 3Gm_3\frac{1}{R^5}(\mathbf{r}, \mathbf{R})\mathbf{R} - Gm_3\frac{1}{R^3}\mathbf{r}. \tag{4.3}$$

First, we consider the first order perturbation. In this case, $\mathbf{r}$ and $\mathbf{R}$ in the right-hand side of equations (4.1), (4.2) can be treated as initial circular and parabolic orbits. The visible singularity $e = 0$ in equation (4.2) is not a problem, since $(\mathbf{r}.\dot{\mathbf{r}})$ contains a factor of e. We will derive the expression for ΔE, which, in the first approximation satisfies

$$\dot{E} = Gm_3\frac{3}{R^5}(\mathbf{r}, \mathbf{R})(\dot{\mathbf{r}}, \mathbf{R}). \tag{4.4}$$

Substitute

$$\mathbf{r} = f(t)\mathbf{r}_0 + g(t)\dot{\mathbf{r}}_0 \tag{4.5}$$

$$\mathbf{R} = Rcos(f_3)\mathbf{i} + Rsin(f_3)\mathbf{j} \tag{4.6}$$

where $\mathbf{r}_0$ and $\dot{\mathbf{r}}_0$ are $\mathbf{r}$ and $\dot{\mathbf{r}}$ at $t = 0$, f_3 is the true anomaly of M_3 and $\mathbf{i}$, $\mathbf{j}$ are unit vectors along the directions of $f_3 = 0°$ and $f_3 = 90°$, respectively.

For a circular binary we have

$$f(t) = cos(nt), \qquad g(t) = \frac{1}{n}sin(nt) \tag{4.7}$$

where n is the mean motion of the binary orbit.

Define the angles

$$cos\theta_1 = \frac{1}{r_0}(\mathbf{r}_0, \mathbf{i})$$

$$cos\theta_2 = \frac{1}{r_0}(\mathbf{r}_0, \mathbf{j}) \tag{4.8}$$

$$cos\theta_3 = \frac{1}{nr_0}(\dot{\mathbf{r}}_0, \mathbf{i})$$

and

$$cos\theta_4 = \frac{1}{nr_0}(\dot{\mathbf{r}}_0, \mathbf{j})$$

Note that the quadrature of equation (4.4) will go from $t = -\infty$ to $+\infty$, so that the quadrature of any odd function of t is zero. Then only even functions of t are kept in the following expression:

$$(\mathbf{r}, \mathbf{R})(\dot{\mathbf{r}}, \mathbf{R}) = na^2R^2[\cos(2nt)\cos^2 f_3 \cos\theta_1 \cos\theta_3 + \cos(2nt)\sin^2 f_3 \cos\theta_2 \cos\theta_4$$
$$+ \sin(2nt)\cos f_3 \sin f_3.(\cos\theta_3 \cos\theta_4 - \cos\theta_1 \cos\theta_2)]$$

With the help of formulae of spherical triangles, we have

$$cos\theta_1 cos\theta_3 = \frac{1}{4}sin^2 i sin(2\Omega) + \frac{1}{2}cos^4(\frac{i}{2})sin2(\Omega+\omega) + \frac{1}{2}sin^4\frac{i}{2}sin2(\Omega-\omega)$$

$$cos\theta_2 cos\theta_4 = \frac{1}{4}sin^2 i sin(2\Omega) - \frac{1}{2}cos^4(\frac{i}{2})sin2(\Omega+\omega) - \frac{1}{2}sin^4\frac{i}{2}sin2(\Omega-\omega)$$

$$cos\theta_3 cos\theta_4 - cos\theta_1 cos\theta_2 = cos^4(\frac{i}{2})sin2(\Omega+\omega) - sin^4\frac{i}{2}sin2(\Omega-\omega)$$

4.2.2 Analytic expressions

The expression for the change in energy E has been obtained by Huang and Valtonen (1987). We proceed in the same manner to obtain the expression for the change in angular momentum. Considering $K = \frac{n}{n_3}$, the relation between K and Q is given by

$$K = \left[\frac{2(m_1+m_2)Q^3}{(m_1+m_2+m_3)}\right]^{1/2} \tag{4.9}$$

where K - is the ratio of the mean motion and Q - is the impact parameter.

The expression for change in energy obtained by Huang and Valtonen (1987) is given by

$$\frac{\Delta E}{E_0} = -\frac{3m_3}{(m_1+m_2+m_3)}\frac{f_1(K)}{K}sin2\Phi \tag{4.10}$$

where

$$f_1(K) = (\frac{I_1}{2} + \frac{I_2}{2} - I_3) \tag{4.11}$$

$$I_1 = 4\int_{-\infty}^{\infty} cos\left[\frac{n}{n_3}(\sigma + \frac{1}{3}\sigma^3)\right]\frac{(1-\sigma^2)^2}{(1+\sigma^2)^4}d\sigma \tag{4.12}$$

$$I_2 = 16\int_{-\infty}^{\infty} cos\left[\frac{n}{n_3}(\sigma + \frac{1}{3}\sigma^3)\right]\frac{\sigma^2}{(1+\sigma^2)^4}d\sigma \tag{4.13}$$

$$I_3 = 8\int_{-\infty}^{\infty} sin\left[\frac{n}{n_3}(\sigma + \frac{1}{3}\sigma^3)\right]\frac{\sigma(1-\sigma^2)^2}{(1+\sigma^2)^4}d\sigma \tag{4.14}$$

and

$$\sigma = tan f_3/2 \tag{4.15}$$

We obtain for the change in angular momentum for the direct case $i - 0°$

$$\frac{\Delta\vec{h}}{h_0} = \frac{3m_3}{(m_1+m_2+m_3)}\frac{[f_2(K) + cos(2\Phi + f_3(K))]}{K}\hat{h} \tag{4.16}$$

where

$$f_2(K) = (2I_6 + I_7) \tag{4.17}$$

and

$$f_3(K) = (2I_4 + I_5) \tag{4.18}$$

and

$$I_4 = 4\int_{-\infty}^{\infty} sin\left[\frac{n}{n_3}(\sigma + \frac{1}{3}\sigma^3)\right]\frac{(1-\sigma^2)^2}{(1+\sigma^2)^4}d\sigma \tag{4.19}$$

$$I_5 = 16\int_{-\infty}^{\infty} sin\left[\frac{n}{n_3}(\sigma + \frac{1}{3}\sigma^3)\right]\frac{\sigma^2}{(1+\sigma^2)^4}d\sigma \tag{4.20}$$

$$I_6 = 16\int_{-\infty}^{\infty} cos^2\left[\frac{n}{n_3}(\sigma + \frac{1}{3}\sigma^3)\right]\frac{\sigma(1-\sigma^2)^2}{(1+\sigma^2)^4}d\sigma \tag{4.21}$$

$$I_7 = 16\int_{-\infty}^{\infty} sin^2\left[\frac{n}{n_3}(\sigma + \frac{1}{3}\sigma^3)\right]\frac{\sigma(1-\sigma^2)^2}{(1+\sigma^2)^4}d\sigma \tag{4.22}$$

The expression for the change in angular momentum for the retrograde case $i = 180°$ is given by

$$\frac{\Delta\vec{h}}{h_0} = \frac{3m_3}{(m_1+m_2+m_3)}\frac{[f_4(K)+cos(2\Phi+f_5(K))]}{K}\hat{h} \tag{4.23}$$

where

$$f_4(K) = (2I_8 + I_9) \tag{4.24}$$

and

$$f_5(K) = (2I_{10} + I_{11}) \tag{4.25}$$

and

$$I_8 = \int_{-\infty}^{\infty} sin\left[\frac{n}{n_3}(\sigma - \frac{1}{3}\sigma^3)\right]\frac{(1-\sigma^2)^2}{(1+\sigma^2)^4}d\sigma \tag{4.26}$$

$$I_9 = 4\int_{-\infty}^{\infty} sin\left[\frac{n}{n_3}(\sigma - \frac{1}{3}\sigma^3)\right]\frac{\sigma^2}{(1+\sigma^2)^4}d\sigma \tag{4.27}$$

$$I_{10} = 16\int_{-\infty}^{\infty} cos^2\left[\frac{n}{n_3}(\sigma + \frac{1}{3}\sigma^3)\right]\frac{\sigma(1-\sigma^2)^2}{(1+\sigma^2)^4}d\sigma \tag{4.28}$$

$$I_{11} = 16\int_{-\infty}^{\infty} sin^2\left[\frac{n}{n_3}(\sigma + \frac{1}{3}\sigma^3)\right]\frac{\sigma(1-\sigma^2)^2}{(1+\sigma^2)^4}d\sigma \tag{4.29}$$

Using the relation between the eccentricity (e), energy (E) and angular momentum (l) given by Goldstein (1950).

$$e = \sqrt{1 + \frac{2El^2}{mk^2}} \tag{4.30}$$

with

$$E = -\frac{Gm_1m_2}{2a} \tag{4.31}$$

$$l = \frac{m_1m_2}{m_1+m_2}\sqrt{G(m_1+m_2)a(1-e^2)} \tag{4.32}$$

then the expression for change in eccentricity takes the form

$$\Delta e = \sqrt{\frac{7}{2}\frac{3m_3}{(m_1+m_2+m_3)}} \quad \frac{f_1(K)}{K}sin2\Phi - \frac{[f_2(K) + cos(2\Phi + f_3(K))]}{2K}\Bigg) \tag{4.33}$$

4.2.3 Analytical discussion

Let $m_1 = m_2 = 0.5$ and $m_3 = 1$, radius of the binary equal to 1 and the gravitational constant $G = 1$. We analyze by using the theoretically obtained equation (4.33) the change in eccentricity and compare the same numerically for the first order term obtained by Huang and Valtonen (1987).

It noticed that for the case of direct orbit, the results are agreed for $Q > 2.5$ and in the case of retrograde orbit for $Q > 1.125$.

We therefore conclude that the change in angular momentum due to the first order term is also valid for the same range. It is important to note that for the smaller values of Q the second order term is neglected.

For direct and retrograde cases, we plot change in energy, angular momentum and eccentricity against the phase angle for the values of Q = 2.5, 3 and 4. It is observed that the phase dependence is nearly same for all the three values. Positive values of $\frac{\Delta E}{E_0}$ in the figures represents tidal capture of m_3 by the binary since its initial orbit is a parabola.

In the range of Q studied, $\frac{\Delta E}{E_0}$ is always less than unity. This implies that the disruption of the binary does not occur for $Q > 2.5$ in the case of direct orbit and for $Q > 1.125$ in the case of retrograde orbit.

Also we notice that the change in the case of direct orbit is an order of magnitude greater than that for the retrograde orbit. This can be understood as due to the difference in interaction time which is greater for the direct orbit.

We substitute the masses of the Earth, the Moon and the Sun for m_1, m_2 and m_3 and let $K = 12$, we obtain the changes in energy and angular momentum for the hypothetical situation in which the Sun encounters the Earth-Moon system in an orbit which is parabolic instead of circular with the same angular velocity ratio at closest approach as that in the Sun, Earth Moon system.

It is of interest to compare the secular change in the semi-major axis in this case with the periodic change predicted by an important perturbation term called the variation in lunar theory (Danby, 1962).

For the variation term in the lunar theory, the first order perturbation term gives,

$$\left(\frac{\Delta a}{a_0}\right)_{max} = \frac{3}{2K_2(1-\frac{1}{K})} = 0.011$$

For the parabolic motion of the Sun the first order term gives

$$\left(\frac{\Delta a}{a_0}\right)_{max} = \left(\frac{\Delta E}{E_0}\right)_{max} = \frac{3f_1(K)}{K} = 0.015$$

The value in the parabolic case is slightly larger than that for the circular case. It shows that the analytical expressions obtained can be used over a wide range of masses m_1, m_2 and m_3. In the following figures we plot the dependence of change in energy, angular momentum and eccentricity on the phase angle, where '$\circ$' indicate energy, '$\diamond$' angular momentum and '$+$' eccentricity.

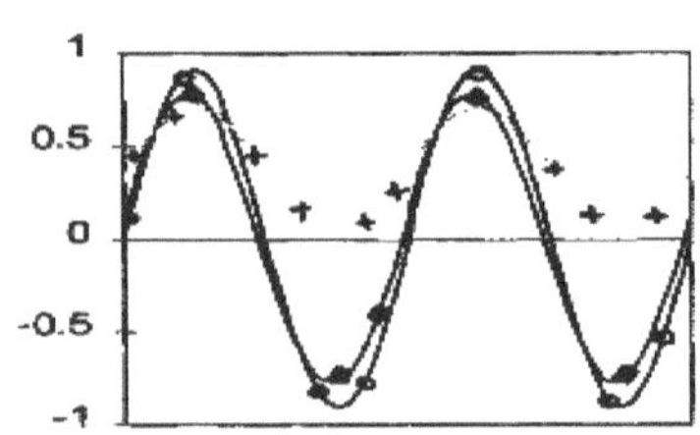

Figure 4.1: (Direct) i = 0 for Q = 2.5 and K = 4

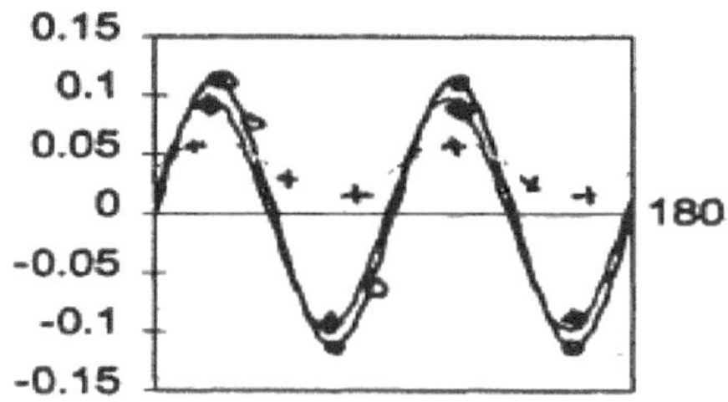

Figure 4.2: (Retrograde) $i = 0$ for $Q = 2.5$ and $K = 4$

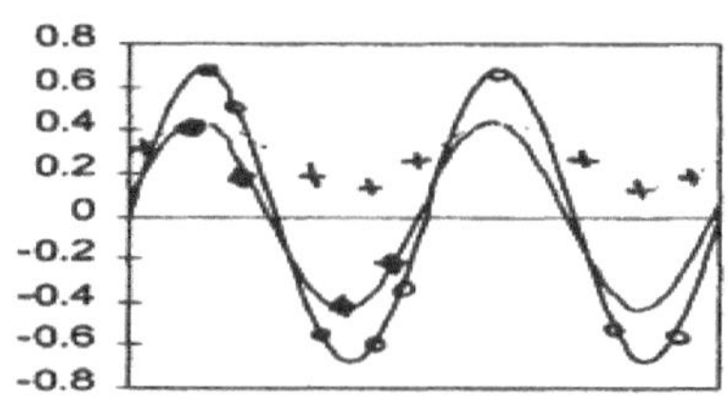

Figure 4.3: (Direct) $i = 0$ for $Q = 3$ and $K = 5.2$

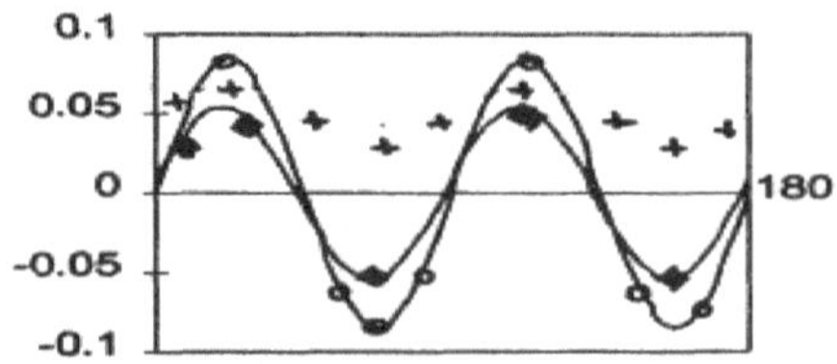

Figure 4.4: (Retrograde) $i = 180$ for $Q = 3$ and $K = 5.2$

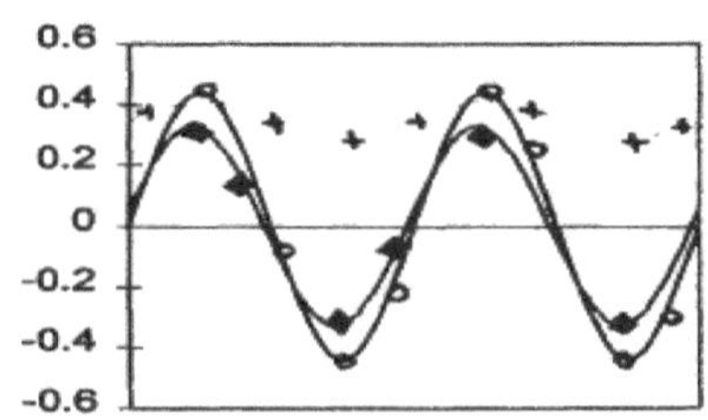

Figure 4.5: (Direct) $i = 0$ for $Q = 4$ and $K = 8$

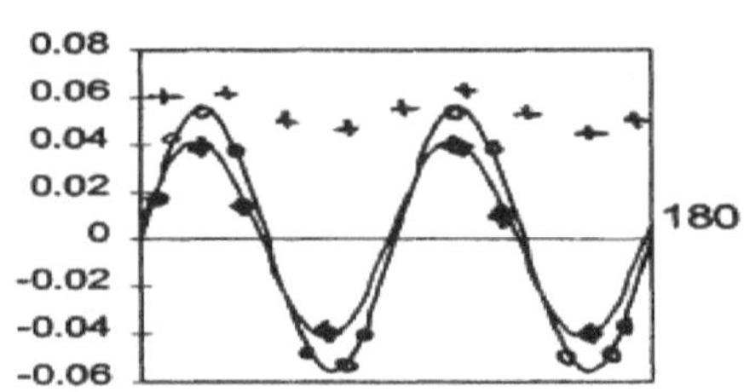

Figure 4.6: (Retrograde) $i = 180$ for $Q = 4$ and $K = 8$

4.3 The effect of parabolic encounter on the eccentricity of binaries

In this section, we discuss the regimes in which the theoretical results are applicable, and to extend them to arbitrary masses and arbitrary geometries. The common assumption is that the encounter is tidal, i.e., $r_p \gg a$, where a is the initial semi-major axis of the binary and r_p is the periastron distance of the encounter and the encounter is slow.

4.3.1 General expressions for the change in eccentricity

Notations, assumptions and basic equations: In this section we specify the notations, assumptions necessary to express the basic equations for the perturbation of a binary by a third body. We consider an encounter between a single star of mass m_3 and a binary whose components m_1 and m_2 and we define $M_{12} = m_1 + m_2$ and $M_{123} = m_1 + m_2 + m_3$. Before the encounter begins, the eccentricity of the binary is e and its semi-major axis is a. The orbit of the third body relative to the barycentre of the binary is hyperbolic with eccentricity is e'. We denote its impact parameter by p and its speed relative to the barycentre by V. Then we have the relation between the eccentricity e' and impact parameter p as discussed by Heggie (1995)

$$e' = \sqrt{1 + \frac{p^2 V^4}{G^2 M_{123}^2}}, \tag{4.34}$$

We obtain results for parabolic encounter by taking $V \to 0$ or $e' \to 1$.

We make two assumptions about the encounter, that it is both *tidal* and *slow*. The encounter *tidal* means, we assume that the closest distance approach of the third body, denoted by r_p, is still considerably larger than semi-major axis a. The ratio of these distances will play the role of a perturbation parameter. The encounter *slow* means, that the angular velocity of the perturber, when it is closest to the binary, is considerably smaller than the angular velocity of the binary components. The equivalent to this condition is given by Heggie (1995)

$$\sqrt{\frac{M_{12}}{M_{123}}}\left(\frac{r_p}{a}\right)^{3/2}\frac{1}{\sqrt{e'+1}} \gg 1, \qquad (4.35)$$

Let **r** be the position of m_2 relative to m_1 and **R** be the position of m_3 relative to the barycentre of the binary. The situation is illustrated in the Figure-4.7.

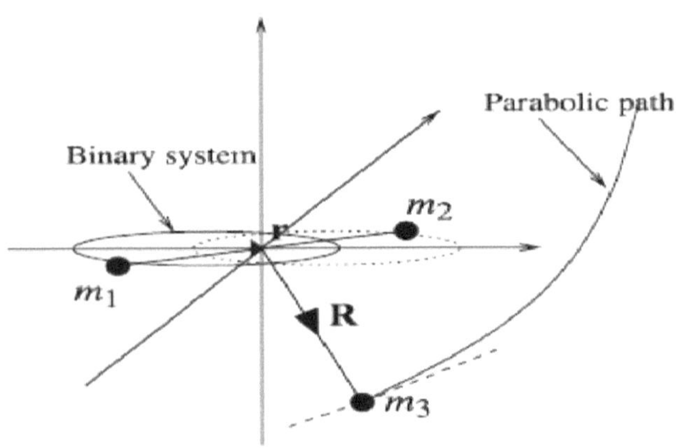

Figure 4.7: Parabolic encounter of a third body with the binary

The equations governing the relative motion of these three stars are derived from the equations of motion of the three-body problem, and may be written the relative motion of the binary components in the form

$$\ddot{\mathbf{r}} = \frac{-GM_{12}\mathbf{r}}{r^3} + F, \tag{4.36}$$

where F is the perturbing function given by

$$F = \frac{Gm_3}{R}\sum_{n=0}^{\infty}\frac{m_1^{n-1} - (-m_2)^{n-1}}{M_{12}^{n-1}}\nabla_r\left[\left(\frac{r}{R}\right)^n P_n\left(\frac{\mathbf{r}.\mathbf{R}}{rR}\right)\right], \tag{4.37}$$

where P_n is the nth Legendre polynomial. Expect for a factor -$\frac{m_1 m_2}{M_{12}}$, equation (4.37) is nothing but the gradient of the mutual potential of the components of the binary and the third body and which can be expanded as a multipole in powers of r/R. The first two terms $n = 0$ and $n = 1$ do not make any contribution to F, hence we will not consider them in the present study.

We compute the change in eccentricity of the binary over the entire duration of the encounter with the third body. For this, it is convenient to use the eccentric axis (Pollard, 1976), which is a vector also associated with the names of Laplace, Hamilton, Lenz and Runge (Goldstein, 1980). For unperturbed Keplerian motion it is a constant vector which points along the major axis of the elliptic orbit of the binary components, and its magnitude e is defined by

$$\mathbf{e} = \frac{1}{GM_{12}}\dot{\mathbf{r}} \times (\mathbf{r} \times \dot{\mathbf{r}}) - \frac{\mathbf{r}}{r} \tag{4.38}$$

Using equation (4.36), in the presence of a perturbing acceleration F, we can show that the eccentricity variation is determined by (Burdet, 1967)

$$GM_{12}\dot{\mathbf{e}} = 2(F.\dot{\mathbf{r}})\mathbf{r} - (\mathbf{r}.\dot{\mathbf{r}})F - (F.\mathbf{r})\dot{\mathbf{r}} \tag{4.39}$$

These are the basical equations to derive the expression for the change in eccentricity.

4.3.2 Binaries with non-zero initial eccentricity

We start with most straightforward and general case, that of a binary of arbitrary initial eccentricity e. Since we are assuming that $r_p \gg a$, hence we neglect all but the lowest non-vanishing terms in equation (4.37). It is clear that, this is the quadrupole term $n = 2$ in the interaction between the binary and the third body.

In order to state this result, we introduce standard angles which describe the relative orientation of the orbits. The line of nodes is defined to be the line of intersection of the orbital planes of the third body and the binary. The ascending node is the part of this line at which the third body crosses the plane of the binary in the direction of the angular momentum vector. Let Ω be the longitude of the ascending node, measured in the plane of motion of the binary from the direction of pericentre. Let i be the inclination of the two orbital planes and ω be the longitude of pericentre of the third body, measured in its plane of motion from the ascending node, in the sense of its motion around the binary. Then the expression for change in eccentricity is given by

Heggie et al (1995)

$$\delta e = -\frac{15}{4}\frac{m_3}{\sqrt{M_{12}M_{123}}}\left(\frac{a}{r_p}\right)^{3/2}\frac{e\sqrt{1-e^2}}{(1+e')^{3/2}}$$
$$\times\left\{\sin^2 i\sin 2\Omega\left[\arccos(-1/e')+\sqrt{e'^2-1}\right]+\frac{1}{3}\left[(1+\cos^2 i)\cos 2\omega\sin 2\Omega+2\cos i\sin 2\omega\cos 2\Omega\right]\frac{(e'^2-1)^{3/2}}{e'^2}\right\} \quad (4.40)$$

However, the result for parabolic motion for the third body can be obtained by taking $e' \to 1$ in the equation (4.40), in this case

$$\delta e = -\frac{15\pi}{16}\left(\frac{2m_3^2a^3}{M_{123}M_{12}r_p^3}\right)^{1/2} e\sqrt{1-e^2}\sin 2\Omega\sin^2 i. \quad (4.41)$$

4.3.3 Initially circular binaries

Secular octupole perturbation: the power-law regime

If the initial eccentricity is zero, the lowest-order results vanishes, which can be observe from the equation (4.40). In this case, for a non-trivial result, it is necessary to include the octupole contribution to the interaction, i.e., the term with $n = 3$ in equation (4.37). The result for the eccentricity induced in an initially circular binary may be stated as follows.

$$\delta e = \frac{15}{8}\frac{m_3|m_1-m_2|}{M_{12}^2}\left(\frac{M_{12}}{M_{123}}\right)^{1/2}\left(\frac{a}{r_p}\right)^{5/2}\frac{1}{e'^3(1+e')^{5/2}}$$
$$\times\left\{\cos^2 i\sin^2\omega\left[f_1(e')(1-\frac{15}{4}\sin^2 i)+\frac{2}{15}(e'^2-1)^{5/2}(1-5\sin^2\omega\sin^2 i)\right]^2\right.$$
$$\left.+\cos^2\omega\left[f_1(e')(1-\frac{5}{4}\sin^2 i)+\frac{2}{15}(e'^2-1)^{5/2}(1-5\sin^2\omega\sin^2 i)\right]^2\right\}^{1/2} \quad (4.42)$$

where

$$f_1(e') = e'^4 \arccos(-1/e') + \frac{\sqrt{e'^2 - 1}}{15}(-2 + 9e'^2 + 8e'^4). \tag{4.43}$$

From the equation (4.42), the change in eccentricity can be obtained for parabolic motion of the third body by taking $e' \to 1$, then we get

$$\delta e = \frac{15\pi}{32}\frac{m_3|m_1 - m_2|}{M_{12}^2}\left(\frac{M_{12}}{2M_{123}}\right)^{1/2}\left(\frac{a}{r_p}\right)^{5/2}\left[\cos^2 i \sin^2 \omega (1 - \frac{15}{4}\sin^2 i)^2 + \cos^2 \omega (1 - \frac{5}{4}\sin^2 i)^2\right]^{1/2}. \tag{4.44}$$

From the equation (4.44), we can notice that the change in eccentricity depends initially on a/r_p. Since the binary is assumed to be circular, and we have averaged over the position of its components. Secondly, the results are proportional to $|m_1 - m_2|$ which we can observe for the $n = 3$ term in equation (4.37). Particulary, for a circular binary with unequal masses, the equation (4.42) is the major contribution to the induced eccentricity at sufficiently large r_p. The change in eccentricity is still proportional to a power of a/r_p, hence we called this as the power-law regime, although it is a higher power than in the case of non-zero initial eccentricity.

Non-Secular quadrupole perturbation: the exponential regime

As mentioned above average over the motion of the binary and taking only the quadrupole interaction, then the eccentricity induced in an initially circular binary vanishes. At small r_p, the assumption on which the method of averaging fails, and we reconsider the problem without averaging. Hence the expression for change in eccentricity is as given by Heggie (1995),

$$\delta e = 3\sqrt{2\pi}\frac{m_3 M_{12}^{1/4}}{M_{123}^{5/4}}\left(\frac{r_p}{a}\right)^{3/4}\frac{(e'+1)^{3/4}}{e'^2}exp\left[-\left(\frac{M_{12}}{M_{123}}\right)^{1/2}\left(\frac{r_p}{a}\right)^{3/2}\frac{\sqrt{e'^2-1}-\arccos(1/e')}{(e'-1)^{3/2}}\right]$$
$$\times\cos^2\frac{i}{2}\left[\cos^4\frac{i}{2}+\frac{4}{9}\sin^4\frac{i}{2}+\frac{4}{3}\cos^2\frac{i}{2}\sin^2\frac{i}{2}\cos(4\omega+2\Omega)\right] \quad (4.45)$$

In the equation (4.45), Ω is measured in the plane of motion of the binary, from its position at the time of closest approach of the third body. And also, this includes a dependence on the phase of the binary. We can simplify further for parabolic motion of the third body by taking $e' \to 1$,

$$\delta e = 3\sqrt{2\pi}\frac{m_3 M_{12}^{1/4}}{M_{123}^{5/4}}\left(\frac{2r_p}{a}\right)^{3/4} exp\left[-\frac{2}{3}\left(\frac{2M_{12}}{M_{123}}\right)^{1/2}\left(\frac{r_p}{a}\right)^{3/2}\right]$$
$$\times\cos^2\frac{i}{2}\left[\cos^4\frac{i}{2}+\frac{4}{9}\sin^4\frac{i}{2}+\frac{4}{3}\cos^2\frac{i}{2}\sin^2\frac{i}{2}\cos(4\omega+2\Omega)\right] \quad (4.46)$$

From equations (4.45) and (4.46), we can notice that the eccentricity is an adiabatic invariant. Usually, this term is applied to a dynamical quantity whose variation is small when the system is subject to a large but slow perturbation. Here, the perturbation is not only slow but small. Also, the exponential form of equations (4.45) and (4.46) applies only to a small set of orbits in which the initial eccentricity vanishes, whereas adiabatic invariants have their special properties more globally. Therefore, we mention this result as the exponential formula, and when it is dominant, called as exponential regime.

4.3.4 Numerical illustrations

We performed the numerical integrations of encounters between a binary and third star. To establish the accuracy of analytical expressions we illustrate the numerical results for the different possible regimes.

Dependence on the periastron distance

We consider the three cases where one of the three expressions (4.41), (4.44) and (4.46) dominates over the entire range of periastron separations, where δe remains small enough that a perturbative calculations applies.

We observe from Figure - 4.8 that, the change in eccentricity of a binary with non-zero initial eccentricity, following an encounter with a star on a general inclined orbit. The three masses are assumed to be equal. We notice that, as long as the three masses are comparable, the magnitude of the fractional change in eccentricity $\delta e/e$ predicted by equation (4.41) is always very small. In the case of initially circular binaries, to isolate the octupole effect for the equation (4.44), we consider a binary with $m_1 \neq m_2$ and we focus on distant encounters with $r_p/a \geq 5$. The variation of the induced eccentricity δe with r_p is shown in Figure - 4.9 for a coplanar, prograde encounter, and for different values of m_2. Notice the power-law behaviour $(r_p^{-5/2})$ compared with the case of an initially non-circular binary $(r_p^{-3/2})$. Also notice, the large drop in δe when m_2 becomes nearly to m_1.

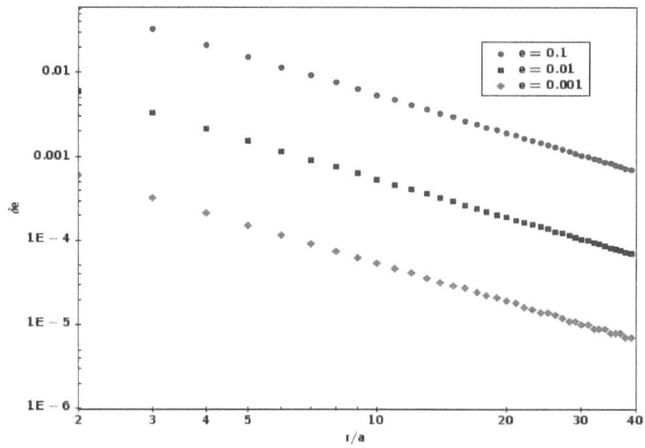

Figure 4.8: Change in eccentricity for an initially non-circular binary (e = 0.1, 0.01, 0.001) perturbed by an encounter with a third body on an inclined, parabolic orbit with periastron distance r.

Angular and phase dependences

We consider the dependence on inclination for the induced eccentricity in the power-law regime. Figure - 4.10, shows the results for $r_p/a = 30$ and for two particular values of ω. For $\omega = 0$, the first term in the angular-dependent factor of equation (4.44) vanishes and the dependence on inclination is $\propto 1 - (5/4)\sin^2 i$, giving zeros of δe at $i \simeq 0.352\pi$ and $i \simeq 0.648\pi$. The maximum δe is obtained for $i = 0$ and $i = \pi$, with a secondary maximum at $i = \pi/2$. For $\omega = \pi/2$, the second term in the angular-dependent factor vanishes and $\delta e \propto \cos i[1 - (15/4)\sin^2 i]$, giving zeros at $i = \pi/2$, $i \simeq 0.173\pi$, and $i \simeq 0.827\pi$. Again the maximum δe is obtained for $i = 0$ and $i = \pi$, with two secondary maxima at $i \simeq 0.335\pi$ and $i \simeq 0.665\pi$.

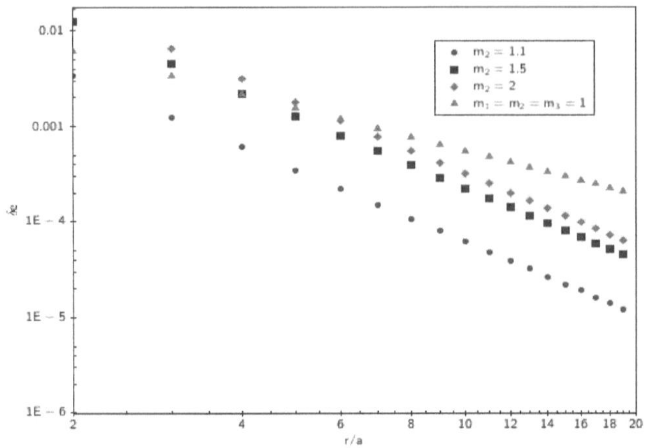

Figure 4.9: Change in eccentricity for a binary with initial e = 0 perturbed by an encounter with a third body on a parabolic orbit with periastron distance r for $m_2 = 1.1, 1.5, 2$. The triangles shows, for comparision, the variation of eccentricity for a non-circular binary with initial e = 0.01.

Dependence on the masses

When the three masses are of comparable magnitude, and $m_1 \neq m_2$, the mass dependence of the change in eccentricity δe is weak. For the various mass values of the third body m_3, the change in eccentricity is as shown in the Figure - 4.11.

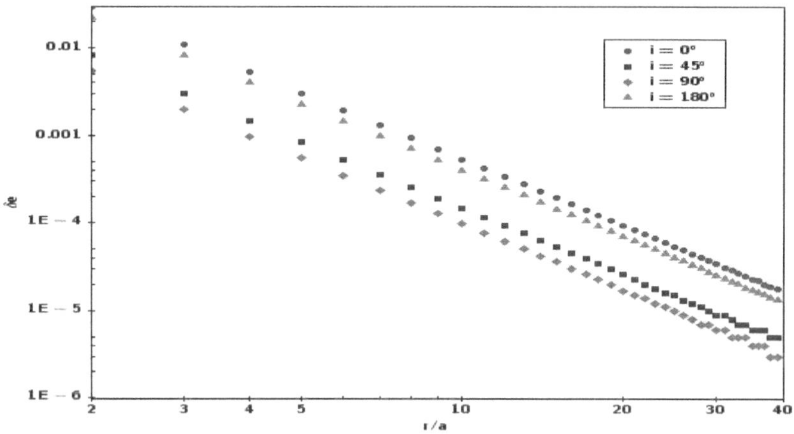

Figure 4.10: Change in eccentricity for an initially circular binary (i = 0, 45, 90, 180) perturbed by an encounter with a third body on a parabolic orbit with periastron distance r (for i = 0 Direct case and for i = 180 Retrograde case).

4.4 Conclusions

In this chapter, we have been discussed analytical expressions for change in angular momentum for the effect of encounter in a circular binary to the direct (orbital inclination $i = 0°$) case and retrograde (orbital inclination $i = 180°$) case. We also given numerical illustrations for the effect of encounters on the eccentricity of non - circular binaries in hyperbolic encounter.

We plot the change in energy, angular momentum and eccentricity against the phase angle for direct and retrograde case for the values of $Q = 2.5$, 3 and 4. It is observed that the phase dependence is nearly same for all the three values. Positive values of $\frac{\Delta E}{E_0}$ denotes tidal capture of m_3

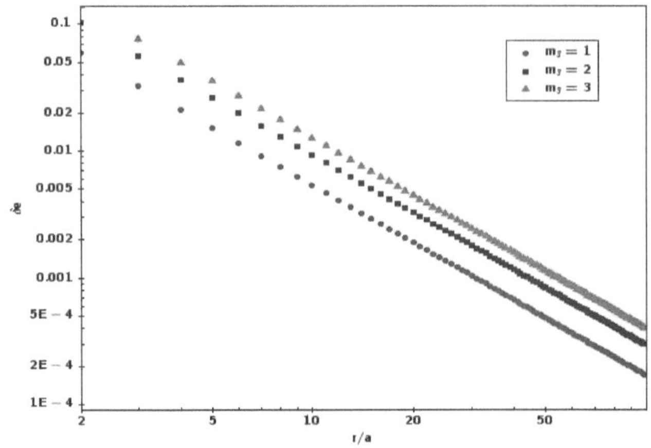

Figure 4.11: Change in eccentricity for a binary with initial $e = 0$ perturbed by an encounter with a third body for $m_3 = 1, 2, 3$ on a parabolic orbit with periastron distance r.

by the binary, since its initial orbit is a parabola.

For all values of Q between 2.5 to 4, $\frac{\Delta E}{E_0}$ is always less than unity. This implies that the disruption of the binary does not occur for $Q > 2.5$ in the case of direct orbit and for $Q > 1.125$ in the case of retrograde orbit.

Also we notice that the change in angular momentum, eccentricity and energy in case of direct orbit is an order of magnitude greater than that for the retrograde orbit. This can be understood as due to the difference in interaction time which is greater for the direct orbit.

The change in the eccentricity δe of a binary of semi-major axis a which experiences an encounter with a third body, if r_p is the distance of closest approach of the perturber which is considerably exceeds a and that the time-scale of the encounter considerably exceeds the period of the binary. With these assumptions our results are taken for the orbit of the third body is nearly parabolic.

When the change in eccentricity, δe, is small compared with the initial eccentricity of the binary, then δe falls of as $(a/r_p)^{3/2}$. The results are shown numerically for a parabolic encounter. These results are affecting by other factors like the masses of the participants, the geometry of the encounter, the initial eccentricity of the binary and the orbit of the perturber. The results comes from the quadrupole interaction between the binary and the third body.

When δe is large compared with the initial eccentricity, and r_p much exceeds a, the final eccentricity falls as $(a/r_p)^{5/2}$, and also depends on various other factors. The results are relevant to binaries whose initial eccentricity is small (i.e., initially circular orbits) and comes from the octupole interaction between the binary and the third body.

Chapter 5

The Scattering of Oort's cloud objects due to encounter with Giant Molecular Clouds

5.1 Introduction

The Oort Cloud (OC) is a massive spherical cloud, the size of this cloud is disputed by different astronomers. Some estimate it to begin at 2000AU and others at 5000AU (AU - Astronomical unit equals to the distance between the Earth and the Sun approximately 150,000,000 km) and ends at 50,000 AU. Others think that it may extend to over 100,000 AU, which would mean its edge would extend to nearly the end of the Solar System. The sphere was named after the astronomer Jan Oort who hypothesized its existence in 1950. The OC is filled with icy objects composed of ammonia, water and methane. Most of the long period comets are believed to come from the OC. In fact, the OC is composed of the same substances as that of comets. The comets are formed when a passing Giant Molecular Cloud (GMC) perturbs some of the objects in the OC out of their orbits. The OC structure is shown in the following Figure - 5.1.

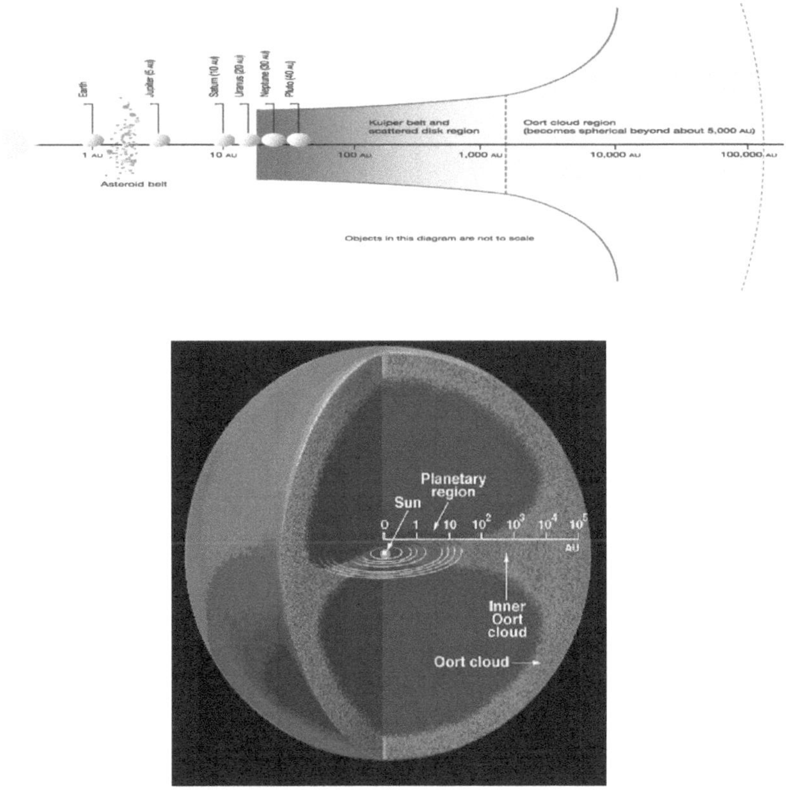

Figure 5.1: Structure of Oort-cloud

In addition to the perturbation by the Galactic tide, comet trajectories are influenced by the nearly passing outer stars and interstellar matter concentrated into the interstellar clouds. Among these interstellar clouds, giant molecular clouds (GMC) have been, particularly, assumed to be the objects that could significantly perturb the OC comets. The outer part of OC can considerably be depleted due to the collisions with interstellar clouds. Works of Napier and Staniucha (1982)

& Clube and Napier (1984), shows that the outer region of the OC, beyond $10000AU$, has been completely destroyed when the Solar System (SS) encountered the GMCs or their substructures. Mazeeva (2004) demonstrated a large influence of the GMCs on the OC comet dynamics.

In the present study, we try to understand an impact of a GMC gravity on the different regions of the OC. We use the FORTRAN code of Jakubik and Neslusan (2009) and adapt it for our simulations. We study the effect of perturbation of GMC on OC comets at varying distances from the Sun.

5.2 Giant Molecular Cloud

Molecular clouds have dense regions where the gas is primarily molecular. Giant molecular clouds (GMCs) are large clouds with $10^4 M_{\odot} < M < 6 \times 10^6 M_{\odot}$ sizes in the range 10 - 100pc. An image of a GMC can be seen in the Figure - 5.2.

The most common molecule in GMCs is molecular hydrogen (H_2), they also contain helium ($\sim 36\%$ by mass) and dust ($\sim 1\%$). In the Milky Way and other galaxies, these are gravitationally bound, discrete structures with masses of the order of $10^4 M_{\odot}$ to $10^6 M_{\odot}$ (Blitz et al. 1980; Fukui et al. 1999; Engargiola et al. 2003). The mass distribution of GMCs is observed to follow a power-law of the form $\frac{dN}{dM} \propto (\frac{M}{M_0})^{\alpha}$, where N is the number of clouds and M_0 is the upper mass limit, which is about $3 - 6 \times 10^6 M_{\odot}$ in the Galaxy (Williams et al. 1997; Rosolowsky,

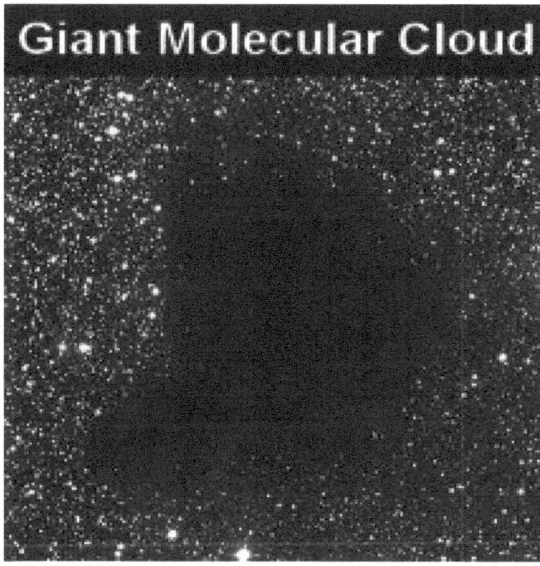

Figure 5.2: Giant - Molecular - Cloud

2005). The index α may be depend on the galactic environment. In the Galaxy, $\alpha = -1.6 \pm 0.1$ (Williams et al. 1997; Rosolowsky, 2005; Blitz et al. 2007). In the Large Magellanic Cloud (LMC), $\alpha \approx -1.7$ and in M_{33}, $\alpha \approx -2.5$ (Blitz et al. 2007).

The structure of GMCs are generally characterized by the total mass and effective radius. Rosolowsky (2007) observed GMCs in the nearby galaxy M31 and obtained the radii range from 19 to 92pc and derived masses from $1.5 \times 10^5 M_\odot$ to $7.8 \times 10^5 M_\odot$. Therefore, a spherical distribution of the GMC density, ρ, is assumed and described by relation

$$\rho(p) = \rho_1 \left(\frac{R_1}{p} \right)^{\alpha} \tag{5.1}$$

where p is the cloud-centric distance, R_1 is the maximum radius of the GMC, and ρ_1 is the density at R_1. The standard value of power-index α is 1. Solomon et al.(1987) illustrated that the density distribution law with $\alpha = 2$ also fits well with photometric observations.

The structure of a GMC can be very complex, certain regions in them can be large clouds of gas and dust, the other regions could be made up of hot dense cores, ridges and star forming regions. To perform realistic N- body simulations would be extremely difficult, rather with the present technological capabilities it is not possible to compute the orbits of every test particle (TP) in OC in the whole period of several million year of the interaction time of the passage of the SS through a GMC with all its complex structures.

For the purpose of simulations a simplified model with a smooth, single law, spherical density profile, namely the Plummer model (Plummer, 1911) is used. The Plummer potential of a spherically symmetric distribution, at a distance p from its centre, is given by

$$\phi(p) = \frac{GM_T}{\sqrt{p^2 + \epsilon^2}} \tag{5.2}$$

where M_T is its total mass and ϵ is the Plummer radius. It can be used to scale the rate of how steeply the density of the condensation rises toward its centre. The density distribution is given by

$$\rho(p) = \frac{\rho_0}{(p^2 + \epsilon^2)^{5/2}}. \tag{5.3}$$

which implies that the mass within the sphere of radius p is given by

$$M_p = \int_0^p 4\pi p^2 \rho(p, \epsilon) dp = \frac{M_T p^3}{(p^2 + \epsilon^2)^{3/2}}. \tag{5.4}$$

The constant ρ_0 is replaced by M_T, we obtained that

$$M_T = \int_0^\infty 4\pi p^2 \rho(p, \epsilon) dp = \frac{4\pi}{3} \rho_0 \epsilon^3. \tag{5.5}$$

The equations of solar motion in the cloud-centric coordinate system are

$$\ddot{p}_\odot = -GM_T \frac{p_\odot}{(p_\odot^2 + \epsilon^2)^{3/2}}. \tag{5.6}$$

5.3 Equations of motion

The equations of TP motion in the GMC-center frame are

$$\ddot{\mathbf{p}} = -\frac{GM_\odot}{r^3}\mathbf{r} - \frac{2v_0^2}{p^2}\mathbf{p} \tag{5.7}$$

where $\mathbf{p} = (p_x, p_y, p_z)$, $p = |\mathbf{p}|$, $\mathbf{r} = (x, y, z)$, and $r = |\mathbf{r}|$. The heliocentric radius-vector of the TP, $\mathbf{r}$, is related to its GMC-centric radius-vector, $\mathbf{p}$, as

$$\mathbf{r} = \mathbf{p} - \mathbf{p}_\odot \tag{5.8}$$

Acceleration of the TP with respect to the Sun is given by

$$\ddot{\mathbf{r}} = -\frac{GM_\odot}{r^3}\mathbf{r} + 2v_0^2 \left. \frac{\mathbf{p}_\odot}{p_\odot^2} - \frac{\mathbf{p}}{p^2} \right) \tag{5.9}$$

For an OC modeled as a Plummer sphere we substitute M_p from the equation, (5.4) in the equation, (5.9) and obtain the heliocentric acceleration of the TP by

$$\ddot{\mathbf{r}} = -\frac{GM_\odot \mathbf{r}}{r^3} - GM_T \left[\frac{\mathbf{p}_\odot + \mathbf{r}}{(p_\odot^2 + 2p_\odot \mathbf{r} + r^2 + \epsilon^2)^{3/2}} - \frac{\mathbf{p}_\odot}{(p_\odot^2 + \epsilon^2)^{3/2}} \right] \tag{5.10}$$

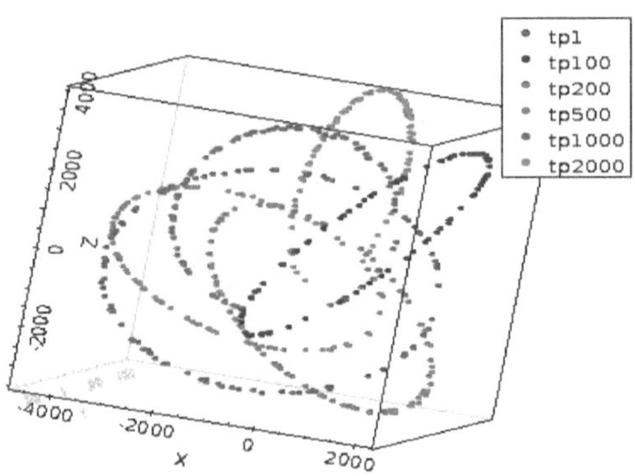

Figure 5.3: Path of the TPs with distance of semi-major axis a = 37000AU with initial inclination i = 0 when the OC perturbed by GMC

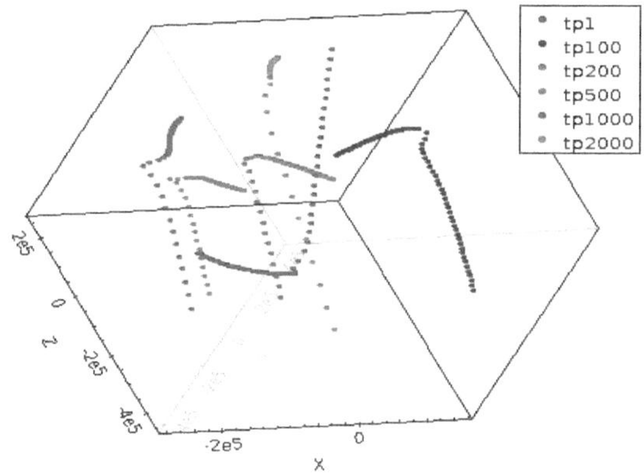

Figure 5.4: Path of the TPs with distance of semi-major axis a = 53000AU with initial inclination i = 0 when the OC perturbed by GMC

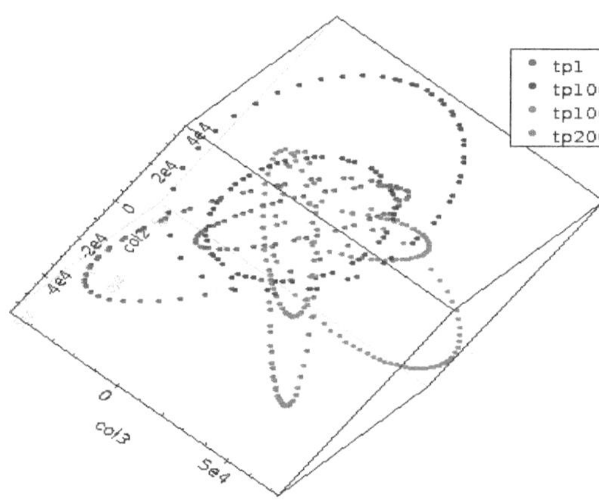

Figure 5.5: Path of the TPs with distance of semi-major axis a = 47000AU with initial inclination i = 0 when the OC perturbed by GMC

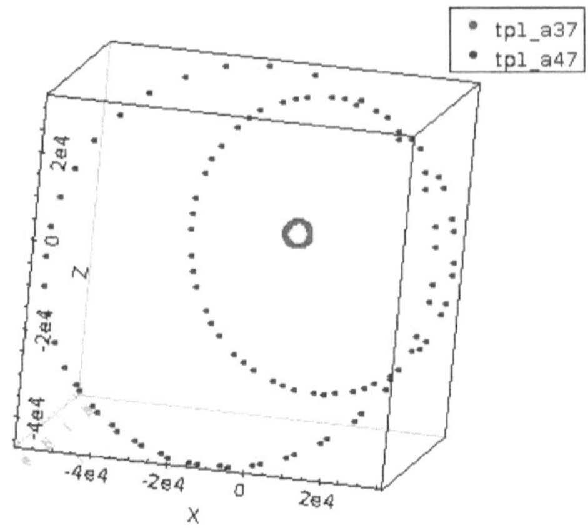

Figure 5.6: Comparison between the path of the TPs with distance of semi-major axis a = 37000AU and a = 47000AU with initial inclination i = 0 when the OC perturbed by GMC

5.4 Description of the simulations

To estimate the effect of perturbation of GMC on the OC and to study the change of OC structure, we solve the equations of motion of TP's using the FORTRAN code provided by Jakubik and Neslusan (2009). We use Plummer model to represent OC, composed of 10^5 TPs and perform simulations and study the effects on the orbits of TPs at varying distance from the Sun in terms of semi-major axis a, varying inclinations with respect to the ecliptic. In this chapter, we present the results for simulation performed for the test particles with distance of semi major axis $a = 37000AU, 47000AU, 53000AU$ with varying inclination i and GMC passage at 100000AU.

Figures - 5.3 - 5.5 give the orbits of TP with a = 37000AU, 47000AU and 53000AU respectively. Figure - 5.6 shows the orbits of the TPs with distance of semi-major axis $a = 37000$AU and $a = 47000$AU and initial inclination $i = 0$ for comparison. Similarly, change in semi-major axis a and change in q with respect to time for the test particles for initial semi major axis $a = 37000$AU is as shown in Figure - 5.7.

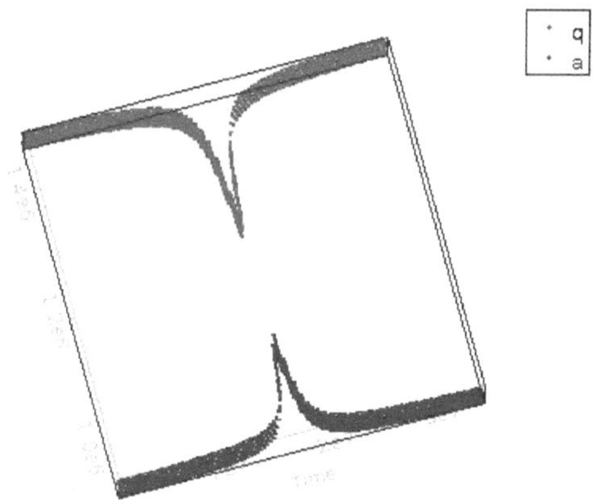

Figure 5.7: Change in semi-major axis a and change in q with respect to time for the TPs for initial semi-major axis a = 37000AU

5.5 Conclusions

It can be noticed from Figure (5.3) that the test particles with semi-major axis $a = 37000$AU and varying i are bound with solar system and are not perturbed by GMC.

The test particles with semi major axis $a = 53000$AU and varying i get strongly perturbed by the GMC and scatter away from the OC. Depleting the density of the outer layer.

The test particles with semi major axis $a = 47000$AU with varying i fall inwards towards the solar system and form long period comets.

The change in semi major axis a (represented in blue) and change in the impact parameter q (represented in red) with respect to time for the test particles are shown in Figure (5.7).

Chapter 6

Scattering phenomenon in a star cluster due to an encounter with interstellar cloud

6.1 Introduction

In this chapter we study star-star encounters and apply the results to study the effect of a passing interstellar cloud on a star cluster. We obtain an expression for the disruption time (t_d) for a star cluster.

6.2 Star - Star Encounter

Let a star of mass m_1 at $\vec{r_1}$ approach a star of mass m_2 at $\vec{r_2}$, where $\vec{r_1}$ and $\vec{r_2}$ are referred to an inertial frame of reference. The equations of motion governing the two stars are:

$$\ddot{\vec{r_1}} = -Gm_2 \frac{\vec{r_1} - \vec{r_2}}{|\vec{r_1} - \vec{r_2}|^3} \tag{6.1}$$

$$\ddot{\vec{r_2}} = -Gm_1 \frac{\vec{r_2} - \vec{r_1}}{|\vec{r_2} - \vec{r_1}|^3} \tag{6.2}$$

By subtraction

$$\ddot{\vec{r}}_1 - \ddot{\vec{r}}_2 = -G(m_1 + m_2)\frac{(\vec{r_1} - \vec{r_2})}{|\vec{r}_1 - \vec{r}_2|^3}$$

Let $\vec{r} = \vec{r}_1 - \vec{r}_2$

Then

$$\ddot{\vec{r}} = -G\frac{(m_1 + m_2)}{r^3}\vec{r} \tag{6.3}$$

is the equation of motion of a star of mass m_1 with respect to the star of mass m_2. If m_1 approaches m_2 from infinite separation with non-zero velocity, the solution of this equation is a hyperbola (Chandrasekhar, 1942) given by

$$\frac{l}{r} = 1 + ecos\theta, \qquad l = a(e^2 - 1) \tag{6.4}$$

$$v_{hyp} = \sqrt{\mu(\frac{2}{r} - \frac{1}{a})}$$

$$v_\infty = \sqrt{\frac{\mu}{-a}}$$

Here, v_{hyp} is hyperbolic velocity and v_∞ is velocity at infinite of a star.

where

$$e = \sqrt{1 + \frac{p^2V^4}{G^2(m_1 + m_2)^2}} \tag{6.5}$$

is the eccentricity of the orbit, l is the semi-latus rectum, θ is the angle measured from the distance of closest approach of m_1, p is the impact parameter and V is the magnitude of the velocity at infinite separation of the two stars. The scattering angles for star-star encounter is as shown in Figure - 6.1.

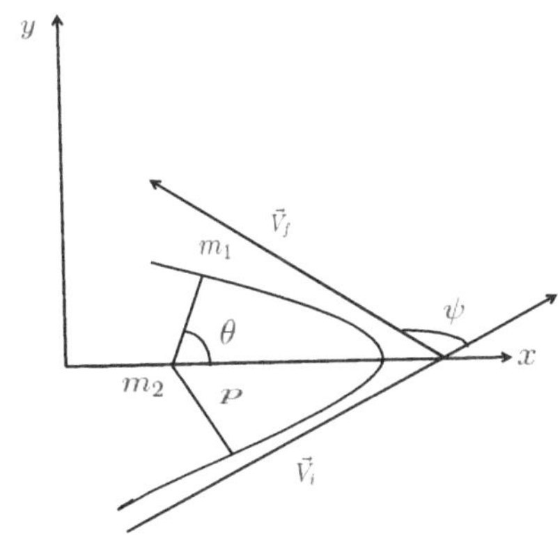

Figure 6.1: Star-Star encounter - Scattering angle (ψ)

Let 2ψ be the angle between the asymptotes and ψ the scattering angle, i.e. the angle through which m_1 is deflected due to the encounter. Then as r tends to infinity, $cos\theta = -\frac{1}{e}$

Since $\psi = 180° - \theta$, $cos\psi = \frac{1}{e}$

$= \Pi - 2$

Therefore

$$tan(\frac{}{2}) = tan(\frac{\Pi}{2} - \psi)$$

$$= cot\psi = \frac{1}{\sqrt{e^2 - 1}} \qquad (6.6)$$

When $e \to 1$, $\psi \to 180°$ (Parabolic orbit)

When $e \to \infty$, $\psi \to 0°$ (Rectilinear orbit)

From equations, (6.5) and (6.6), it follows that

$$\tan\frac{}{2} = \frac{G(m_1 + m_2)}{pV^2} \qquad (6.7)$$

The change in velocity of the two stars with respect to their centre of mass due to the encounter:

Let m_2 be taken as the origin of the coordinate system. Choose the coordinate axes so that X is in the direction of m_1 when the latter is at its closet approach and XY is in the orbital plane.

The magnitude of the relative velocity is the same before and after the encounter. Hence

$$|\vec{V_i}| = |\vec{V_f}| = V$$

Where $|\vec{V_i}|$ and $|\vec{V_f}|$ are the initial and final relative velocities of m_1 with respect to m_2

Then

$$\vec{V_i} = (+Vcos\psi, +Vsin\psi, 0)$$

$$\vec{V_f} = (-V cos\psi, +V sin\psi, 0)$$

$$\Delta\vec{V} = \vec{V_f} - \vec{V_i} = (-2V cos\psi, 0, 0)$$

Therefore $\Delta\vec{V}$ is in the x-direction, i.e. m_1 acquires an increment in velocity in the direction of m_2 and its magnitude is given by

$$(\Delta V)_{m_1} + (\Delta V)_{m_2} = |\vec{\Delta}V| = 2V cos\psi = 2V\frac{1}{e}$$

$$= \frac{2VG(m_1 + m_2)}{\sqrt{G^2(m_1 + m_2)^2 + p^2V^4}} \tag{6.8}$$

The change in velocity of m_1 with respect to the centre of mass is

$$(\Delta V)_{m_1} = \frac{m_2}{m_1 + m_2}\Delta V = \frac{2VGm_2}{\sqrt{G^2(m_1 + m_2)^2 + p^2V^4}} \tag{6.9}$$

The change in velocity of m_2 with respect to the centre of mass is

$$(\Delta V)_{m_2} = \frac{m_1}{m_1 + m_2}\Delta V = \frac{2Vm_1}{\sqrt{G^2(m_1 + m_2)^2 + p^2V^4}} \tag{6.10}$$

Note the dependence on masses. Also note that since

$$tan\frac{\ }{2} = \frac{G(m_1 + m_2)}{pV^2} \tag{6.11}$$

$$sin\frac{}{2} = \frac{\frac{G(m_1+m_2)}{pV^2}}{\sqrt{1+\frac{G^2(m_1+m_2)^2}{p^2V^4}}}$$

$$= \frac{G(m_1+m_2)}{\sqrt{p^2V^4+G^2(m_1+m_2)^2}}$$

Comparing this with (6.8), we get

$$\frac{\Delta V}{V} = 2sin(\psi/2)$$

Also

$$\frac{(\Delta V)_{m_1}}{V_{m_1}} = \frac{\left(\frac{m_2}{m_1+m_2}\right)\Delta V}{\left(\frac{m_2}{m_1+m_2}\right)V} = \frac{\Delta V}{V} = 2sin(\psi/2) \tag{6.12}$$

Where $(V)_{m_1}$ denotes the speed of m_1 with respect to the centre of mass of m_1 and m_2.

6.3 Tidal effects of passing interstellar clouds on star clusters

Spitzer (1958) showed that the tidal field of the passing interstellar clouds has important effects on the dynamical evolution of a galactic cluster. The tidal field leads to the acceleration of the stars in a cluster, increases the total cluster energy and leads to the expansion and ultimate disruption of the cluster. This may be seen from the following considerations.

Let $\Delta\vec{v}$ be the change in velocity of a star in a cluster with respect to the centre of mass of the cluster due to an encounter with an interstellar cloud. Then the change in kinetic energy of the star is

$$\frac{1}{2}m\left[(\vec{v}_i + \Delta\vec{v})^2 - \vec{v}_i^2\right] = m\left[\vec{v}_i\Delta\vec{v} + \frac{1}{2}(\Delta v_i)^2\right] \tag{6.13}$$

where m is the mass and $\vec{v}_i$ is the initial velocity of the star. When we consider a large number of stars, we may set statistically $\vec{v}_i.\Delta\vec{v} = 0$. So that the change in the kinetic energy of the whole cluster is

$$\Delta T = \frac{1}{2}\sum m(\Delta\vec{v})^2 \tag{6.14}$$

where the summation is over all the stars in the cluster. Most of the change in the kinetic energy of the cluster would take place when the cloud is closest to the cluster. The tidal field is thus of impulsive nature and without much loss of accuracy it may be assumed that during the time the cluster gains kinetic energy, the change in the mass distribution of the cluster is negligible, so that the change in the potential energy of the cluster during the encounter may be ignored. It thus follows that $\Delta T = \Delta U$, where U is the total energy of the cluster. Since $(\Delta\vec{v})^2$ is a positive quantity and U is negative for a stable system, it follows that U would be less negative after the encounter in the new equilibrium established after the encounter. The potential energy $\Omega = \frac{1}{2}U$ by Virial theorem. Therefore Ω would also be less negative than before. Since

$$\Omega = -\frac{GM^2}{2\bar{R}}$$

it follows that $\frac{GM^2}{2\bar{R}}$ becomes smaller as a result of the encounter. Thus $\bar{R}$ increases, that is, the cluster expands.

Let us proceed to compute the rate at which U increases following the analysis of Spitzer (1958). We shall assume that the cloud passes by the cluster sufficiently rapidly so that the change in velocity of each star may be computed as though the stars were not moving. We define a coordinate system with origin at the centre of the cluster and with X-axis pointing towards the cloud when it is at its distance p of closest approach to the cluster centre. The Y-axis is taken parallel to the velocity of the cloud when it is at its closest approach. The Z-axis is taken perpendicular to the orbital plane.

The tidal acceleration $\vec{F}$ is most conveniently expressed in a rotating reference frame, ξ, η, ζ in which ξ points to the instantaneous position of the cloud, η is perpendicular to ξ in the orbital plane, and ζ is parallel to z. Consider a star at point P in the cluster and let $\vec{r}$ be its radius vector with respect to the cluster center O. Let $\vec{r}'$ and $\vec{\Delta}$ be the instantaneous radius vectors of the cloud from the cluster centre, and from P respectively. Let m, M and M' be the masses of star, the cluster and the cloud respectively. The geometry of the star, cluster and cloud is as shown in Figure - 6.2.

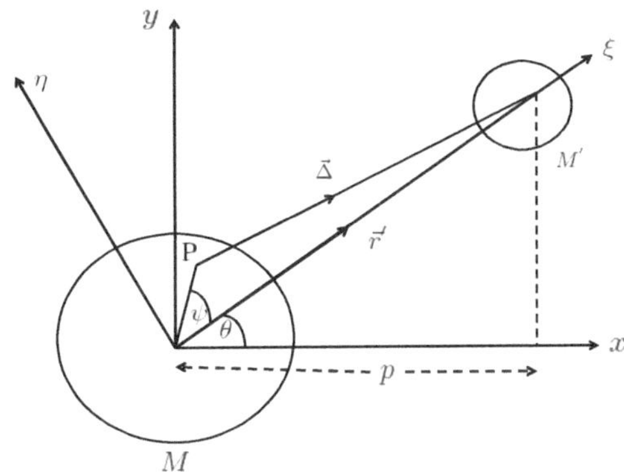

Figure 6.2: The geometry of the star cluster cloud

The potential at P due to the cloud is

$$V_p = -\frac{GM'}{\Delta}$$

$$= -\frac{GM'}{(r^2 + r'^2 - 2rr' cos\psi)^{1/2}}$$

$$= -\frac{GM'}{r}\left(1 - \frac{2r}{r'}cos\psi + \frac{r^2}{r'^2}\right)^{-1/2} \tag{6.15}$$

Where ψ is the angle between r and r'.

Expanding the right hand side by binomial theorem or Legendre polynomials, we get

$$V_p = -\frac{GM'}{r'} - \frac{GM' rcos}{r'^2} - \frac{GM' r^2(3cos^2\psi - 1)}{2r'^3} \tag{6.16}$$

We shall assume that $r \ll r'$ and neglect further terms. The gravitational force per unit mass at P is obtained by taking the gradient of the potential with respect to the coordinate of P. The first term on the right hand side of equation (6.16) does not contain the coordinates of P and so does not give rise to any force at P. The second term may be written -$\frac{GM'\xi}{r'^2}$ since $\xi = rcos\psi$. We note that the force due to this term is given by

$$(F_\xi)_2 = -\frac{d}{d\xi}(-\frac{GM'}{r'^2}\xi) = \frac{GM'}{r'^2}$$

$$(F_\eta)_2 = -\frac{d}{d\eta}(-\frac{GM'}{r'^2}\eta) = 0$$

Thus the second term gives rise to a uniform force with intensity $\frac{GM'}{r'^2}$. This can be neutralized if the coordinate axes are rotating with an acceleration $\frac{GM'}{r'^2}$ while the centre of mass of the cluster always remains at the origin. The remaining terms constitute the tidal potential V_T since we assume the $r \ll r'$, only the first term in the series of the tidal potential will be considered. Hence

$$V_T = -\frac{GM'}{r'^3}, \qquad \frac{3r^2cos^2\psi - r^2}{2}$$

Now, $\xi = rcos\psi$

$r^2 = \xi^2 + \eta^2 + \zeta^2$

$$V_T = -\frac{GM'}{r'^3}\left(\xi^2 - \frac{1}{2}\eta^2 - \frac{1}{2}\zeta^2\right) \tag{6.17}$$

The tidal acceleration in the rotating coordinate system is therefore given by

$$\vec{F} = -\nabla V_T$$

whence

$$F_\xi = \frac{2GM'\xi}{r'^3}, \quad F_\eta = -\frac{2GM'\eta}{r'^3}, \quad F_\zeta = -\frac{2GM'\zeta}{r'^3} \tag{6.18}$$

Suppose we reckon time from the instant of closest approach. Let V be the speed of the cloud which we shall assume to be constant. Then from geometry, as shown if Figure - 6.3, we have

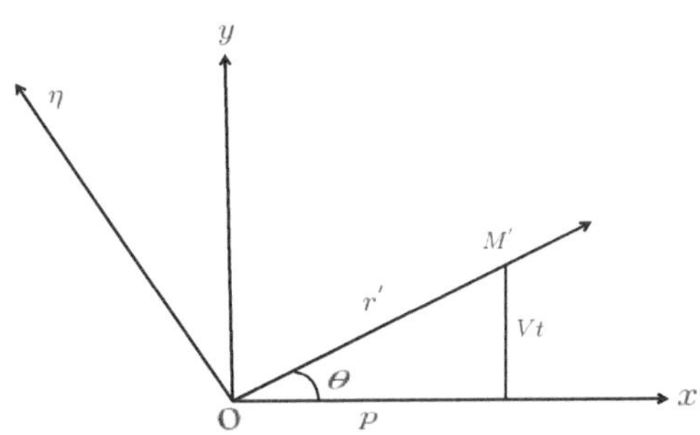

Figure 6.3: The geometry of the rotating co-ordinate frame

$$r'^2 = p^2 + V^2t^2 \tag{6.19}$$

The rotating coordinate frame is related to the fixed frame by the equations

$$\xi = xcos\theta + ysin\theta, \qquad \eta = -xsin\theta + ycos\theta \tag{6.20}$$

Where

$$\tan\theta = \frac{Vt}{p} \tag{6.21}$$

The changes in the velocity components of the star at P as measured in the fixed frame are obtained from the relations

$$\frac{dV_x}{dt} = F_x = F_\xi cos\theta - F_\eta sin\theta$$

$$\frac{dV_y}{dt} = F_y = F_\xi sin\theta + F_\eta cos\theta \tag{6.22}$$

$$\frac{dV_z}{dt} = F_z = F_\zeta$$

Substituting for F_ξ, F_η, F_ζ from equation (6.18) and using equation (6.20)

we obtain

$$\ddot{x} = \frac{dV_x}{dt} = \frac{GM'}{r'^3}\left[x(2 - 3sin^2\theta) + 3ysin\theta cos\theta\right]$$

$$\ddot{y} = \frac{dV_y}{dt} = \frac{GM'}{r'^3}\left[3x sin\theta cos\theta + y(2 - 3cos^2\theta)\right] \tag{6.23}$$

$$\ddot{z} = \frac{dV_z}{dt} = -\frac{GM'}{r'^3}z$$

The change in the velocity due to the encounter is obtained from the equation

$$\Delta\vec{V} = \int_{-\infty}^{+\infty} \frac{d\vec{V}}{dt}\, dt \tag{6.24}$$

We integrate equation (6.23) with the use of equations (6.19) and (6.21) assuming that, x, y, z and v do not change with time (impulsive approximation) we get,

$$\Delta V_x = \frac{2GM'x}{p^2V}, \quad \Delta V_y = 0, \quad \Delta V_z = -\frac{2GM'z}{p^2V} \tag{6.25}$$

It follows that ΔU, the increase in total energy of the cluster, which is equal in our impulsive approximation to ΔT is given by $\Delta U = \frac{1}{2}\sum m(\Delta\vec{V})^2$ where the summation is over all the stars in the cluster.

Using equation (6.25) we get,

$$\Delta U = \frac{1}{2}\sum m \left(\frac{2GM'}{p^2V}\right)^2 (x^2 + z^2) \tag{6.26}$$

Let r_c^2 be the mean square cluster radius defined by

$$Mr_c^2 \equiv \sum m(x^2 + y^2 + z^2)$$

$$\equiv \sum \frac{3}{2} m(x^2 + z^2) \qquad (6.27)$$

Then

$$\Delta U = \frac{4}{3} M \quad \frac{GM' r_c}{p^2 V} \Big)^2 \qquad (6.28)$$

The internal energy U is related to the potential energy Ω of the cluster by Virial theorem. We have

$$U = \frac{\Omega}{2} = -\frac{rGm^2}{2r_c} \qquad (6.29)$$

where, as before, r_c^2 is the mean square value of r for all the cluster stars and γ depends on the density distribution within the cluster. For a homogeneous sphere $\gamma = 0.465$ and for a polytopic sphere of index n, γ varies from 0.469 to 0.564 as n increases from 1 to 4. These values are based on radius of gyration computed for polytropes by Motz (1952), r_c is $0.772R$ for a homogenous sphere and $0.188R$ for a polytropic sphere of index $n = 4$ where R is the radius of the sphere. From equations, (6.28) and (6.29) it follows that the fractional change in the internal energy of the cluster due to a single encounter with a cloud is given by

$$\frac{\Delta U}{|U|} = \frac{8GM'^2 r_c^3}{3rMp^4V^2} \qquad (6.30)$$

Equation (6.28)gives the change in energy of the cluster due to a single encounter with the cloud. The rate of change of U is given by ΔU multiplied, by the number of collisions per unit time. The collision frequency is given by

$$f = \sigma_n V$$

Hence

$$\frac{dU}{dt} = \int_0^{+\infty} V_m(M^{'})dM^{'} \int_{R'}^{+\infty} \Delta U 2\Pi p dp \tag{6.31}$$

where R is the radius of the cloud $n(M^{'})$ is the number of clouds per unit volume with a mass between $M^{'}$ and $M^{'} + dM^{'}$. We shall assume that all clouds have the same speed V. Strictly the lower limit of the second integral should be zero. Encounters for which p is less than $R^{'}$ will increase U somewhat but we neglect them. Substituting for ΔU from equation (6.28) into (6.31), we get

$$\frac{dU}{dt} = \frac{8\Pi G^2 M r^2}{3V} \int_0^{+\infty} n(M^{'})M^{'2}dM^{'} \int_{R'}^{+\infty} \frac{dp}{p^3} \tag{6.32}$$

$$= \frac{8\Pi G^2 m r_c^2}{3V} \int_0^{+\infty} n(M^{'})\frac{M^{'2}}{2R^{'2}}dM^{'} \tag{6.33}$$

To evaluate the equation (6.32) approximately, we see that

$$\frac{M^{'}}{R^{'2}} = \frac{4}{3}\Pi R^{'} \rho^{'}$$

where $\rho^{'}$ is the internal density of the cloud, assumed to be a uniform sphere of radius $R^{'}$. We assume that the product $R^{'}\rho^{'}$ is independent of the size of the cloud which means that we assume an optical extinction per cloud independent of the cloud size. It was pointed out by Bok that clouds seem to have a tendency in this direction. We then get

$$\frac{dU}{dt} = \frac{16\Pi^2 G^2 M r_c^2 R^{'} \rho^{'} \rho_a}{9V} \tag{6.34}$$

where $\rho_a = \int_0^{+\infty} n(M^{'})dM^{'}$ is the average space density of the clouds in the volume of the Galaxy in which the cluster moves. Eliminating r_c from equation (6.34) by means of equation (6.29), we get

$$\frac{dU}{dt} = \frac{4\Pi^2 G^4 M^5 R^{'} \rho^{'} \rho_a}{9VU^2} \tag{6.35}$$

or

$$\int_U^0 U^2 dU = \frac{4\Pi^2 G^4 M^5 R^{'} \rho^{'} \rho_a}{9V} \int_0^{t_d} dt \tag{6.36}$$

where we define the disruption time t_d as time at which U become zero. Performing the integration and then making use of equation (6.29) to eliminate U we get

$$t_d = \frac{r}{8\Pi G} \frac{\rho_c}{\rho_a} \frac{V}{R^{'}\rho^{'}} \tag{6.37}$$

6.4 Conclusions

In this chapter, we consider the effect of an encounter of an interstellar cloud on a star cluster. The tidal field of the passing interstellar cloud has important effects on the dynamical evolution of a star cluster. The tidal field leads to the acceleration of the stars in the cluster, consequently increasing the total cluster energy and resulting in the expansion and ultimate disruption of the cluster. An expression for disruption time is derived.

Chapter 7

Conclusions

7.1 Our work

As part of our work, in **chapter-1** we have considered the three-body problem and formulated the equations of motion. In cartesian coordinates the three body system consists of nine differential equations, each of second order, and hence the order of system is 18. We have shown the reduction of order of the system from 18th to 6th by taking the six integrals of center of mass motion of three-body system, three integrals of angular momentum, the integral of energy, together with the elimination of the time and the elimination of ascending node. So finally the equations of motion may be reduced to a system of the 6th order.

The general n-body problem is unsolvable in closed form for $n \geq 3$, i.e., given the coordinates and velocities of n mass points moving under their own mutual gravitational attractions, one can not predict the motion for any arbitrary succeeding time interval. We have discussed the special configurations of the masses in the three-body problem which do lead to the closed form of solutions. Particular solutions are those solutions in which the geometric configuration of three bodies remains invariant with respect to time. Thus, either the configuration simply rotates in its own plane around the centre of mass, or an expansion or contraction takes place in which the mutual distances between the three bodies remain in fixed ratios to each other. If there is no change in scale, the solutions are called stationary.

Besides these, we have focused on particular solutions collinear and equilateral configurations found by Lagrange. In the collinear case, the bodies are all set in motion from positions on a straight line, and given appropriate initial conditions, they continue to stay on that line while the line rotates in a plane about the centre of mass of three bodies. In the equilateral case, the initial positions of the three bodies are at the vertices of an equilateral triangle, and the bodies continue to move as though attached to the triangle that rotates about the centre of mass.

In **chapter-2**, we studied the dynamics of the Sun - Earth - Moon system with special attention to the effects of Sun's perturbations on the Moon's orbit around the Earth. We have focused on secular effects of the regression of the nodes, the advance of the perigee and the increase in the moon's mean longitude over the period from January 2000 to December 2050, i.e, 630 synodic months. We focused on this aspect of the synodic period variation of Moon and show how the changes in the angular separations of the longitudes of the perihelion of the Earth and the Moon at the time of the new moon are correlated. The amplitude of fluctuation Δp, which is the difference between the actual synodic period and its mean value, undergoes two sinusoidal nature of variations. There is a smaller cycle of about 14 synodic months and a larger cycle of about 112 synodic months. This period is close to the period of one complete revolution of Moon's perigee.

The chosen interval of 630 synodic months covers five larger cycles. The smaller cycle resembles a sine curve while the larger cycle resembles a double sine curve.

We have also studied the effects of the variation of Earth - Sun distance on Δp. It is found to increase up to a maximum value and then decrease, and reaching the maximum lowest value then it starts to increase. The phase difference, at the time of new moon, between the perigee of the Moon and the Sun remains within 2^0 to 4^0. The difference between the longitudes of the Moon and that of it's perigee is about 26^0 at every new moon in the entire cycle. While the difference between the longitudes of the Sun and Sun's perigee lies between 28^0 to 29^0 at every new moon in the cycle.

In **chapter-3**, we have taken Burrau's problem as an application of three - body scattering problem. This can be defined in the following way. The masses of the particles are $m_1 = 3$, $m_2 = 4$ and $m_3 = 5$, these mass points are initially located at the vertices of a right angle triangle with sides 3,4 and 5 so that the corresponding masses and sides are opposite. In this problem, the bodies are assumed to be point masses and the Newtonain equations of motion are applied at closed encounters. We have written a FORTRAN programme and perform simulations and investigate the nature of the solution for large intervals of time. We observed that initially the three bodies interact forming double or triple systems for short time and ultimately the larger two bodies form a stable binary while the smallest body escapes the system.

We also perform simulations with the same code for other triple system such as equal masses placed on vertices of equilateral triangle and observed that the three bodies collide at centriod of triangle. In this chapter we give a proof of the theorem that, a necessary but not sufficient condition for simultaneous triple collision is that the total angular momentum of system is zero. Our numerical simulation illustrates the result. In general a three body system is unstable. In a triple system most of the time the smallest mass escape the system while the larger two form a stable binary. This is observed in star clusters in which a large number of binaries are formed as an end product of triple star encounters. Presence of binaries play a very important role in the dynamical evolution of star clusters.

In **chapter-4**, we consider yet another application of three - body scattering. Here, we investigated a parabolic encounter of a single star with a circular binary and study the effects of the encounter on the eccentricity of binary for the case of binaries with non zero initial eccentricity and initially circular binaries. In this, we have point out the changes in orbital elements i.e., angular momentum, eccentricity etc of the binary caused by an encounter with another star. We extended Huang and Valtonen (1987) analytic treatment to obtain angular momentum transfer and change in eccentricity for the case of a parabolic encounter of a star with a circular binary. We also study the effect of encounters on the eccentricity of non - circular binaries in hyperbolic encounter.

In **chapter-5**, we have taken the Sun - Oort cloud (OC) - Giant Molecular Cloud (GMC) as a triple system. We estimate the effect of perturbation of GMC on the OC and to study the change of OC structure. We solve the equations of all considered test particles (TPs) using the FORTRAN code provided by Jakubik and Neslusan (2009). We model the OC as a Plummer sphere with 10^5 particles and perform simulations and study the effects on the orbits of TPs at varying distance from the Sun in terms of semi-major axis a = 37000AU, 47000AU and 53000AU, varying inclinations with respect to the ecliptic. Also we compared the path of the TPs with distance of semi-major axis a = 37000AU and a = 47000AU with inclination $i = 0$. We observed that the TPs with semi-major axis a = 37000AU and varying inclination i are bound with the solar system and are not perturbed by GMC. The TPs with semi-major axis a = 47000AU with varying i fall inwards towards the solar system and form long period comets. The TPs with semi-major axis a = 53000AU and varying i get strongly perturbed by the GMC and scatter away from the OC. Depleting the density of the outer layer. The change in semi major axis a (represented in blue) and change in the impact parameter q (represented in red) with respect to time for the test particles are shown.

In **chapter-6**, we have considered the effect of an encounter of a passing interstellar cloud on a star cluster. We show that it has important effects on the dynamical evolution of the star cluster. The tidal field leads to the acceleration of the stars in a cluster, increases the total cluster energy and leads to the expansion and ultimate disruption of the cluster. An expression for the disruption time for a cluster is derived.

7.2 Future Work

The extension of this work to the case of binary-binary encounters. If star clusters initially contain large numbers of binaries, such encounters may well be of comparable importance to encounters between single stars, or encounters between single stars and binaries. We can extend to study the interaction of galaxies which often result in the formation of merged system or formation of the galactic tails and bridges.

References

[1] Abyankar, K. D., “Astrophysics of the solar system”, Ch.II, Universities press (India) Ltd., Hyderabad (1999).

[2] Arenstorf, R. F., A new method of perturbation theory and its application to the satellite problem of celestial mechanics, J. Rein Angew. Math., 221, 113 (1966).

[3] Arnold, V. I, Proof of the theorem of A. N. Kolmogorov concerning the conservation of quasi-periodic motions with small change of the Hamiltonian function, Usp. Mat. Nauk, 18, 13 (1963).

[4] Bailey, M.E., Mon. Not. R. astr. Soc.,218, 1 (1986).

[5] Ballabh, G. M. and Alladin, S. M., Bull. Astr. Soc.India, 30, 601 (2002)

[6] Biermann, L., Astron. Papers Dedicated to Bengt Stromgren, Symposium, Copenhagen, 327 (1978).

[7] Binney, J. and Tremaine, S., Galactic Dynamics, Princeton Univ. Press, Princeton (1987).

[8] Birkhoff, G. D., “Recent advances in dynamics” science, 51, 51 (1920).

[9] Bolin, K., “Uber ein zweckmassiges Beispiel der Bewegung im allgemeinen Dreikorper-probleme,” Astronomiska Iakttagelser och Undersokningar a Stockholms Observatorium, 10, No. 11 (1923).

[10] Born, M., "The Mechanics of the Atom", Ungar, New York (1960).

[11] Brasser, R., Duncan, M.J. and Levison, H.F., Icarus 184, 59 (2006).

[12] Brasser, R., Duncan, M.J. and Levison, H.F., Icarus 191, 413 (2007).

[13] Brasser, R., Duncan, M.J. and Levison, H.F., Icarus 196, 274 (2008).

[14] Broucke, R.A., J. Guid. Control Dyn., 26, 27 (2003).

[15] Bulirsch, R., Stoer, J., Num. Math., 8, 1 (1966).

[16] Burdet, C.A. and Angew. Z., Math. Phys., 18, 434 (1967).

[17] Byatt-Smith, J.G.B. and Davie, A.M., Proc. R. Soc. Edin., 114A, 243 (1990).

[18] Camilo, F., Nice, D.J. and Taylor, J.H., ApJ, 412, L37 (1993).

[19] Clube, S.V.M. and Napier, W.M., Mon. Not. R. astr. Soc.,208, 575 (1984).

[20] Danby, J.M.A., " Fundamentals of Celestial Mechanics", Macmillan, New York (1962).

[21] Domingos, R. C., Moraes, R. V. and Prado, A.F.B.A., Math. Probl. Eng, 2008, 763654 (2008).

[22] Dones, L., Weissman, P.R., Levison, H.F. and Duncan, M.J., Comets II, ed., M.C. Festou, H.U. Keller, and H.A. Weaver, Univ. Arizona Press, Arizona, 153 (2004).

[23] Duncan, M., Quinn, T. and Tremaine, S., Astron. J. 94, 1330 (1987).

[24] Dyczynski, P.A., Leto, G., Jakubik, M., Paulech, T. and Neslusan, L., Astron. Astrophys. 487, 345 (2008).

[25] Garcia, F. L., "Extension de la tabla de Zumkley por prolongaciones analiticus approximadas," in Report of the Comision de Investigacion Cientifica, Provincia Buenos Aires, La Plata (1996).

[26] Goldstein H.,"Celestial Mechanics", Addison-Wesley, p. 78 (1950).

[27] Gutzwiller, M. C., Moon-Earth-Sun: The oldest three-body problem, Reviews of Modern Physics,70, No.2, 589, (1998).

[28] Harrington, R.S., Astron. J., 80, 1081(1975).

[29] Harrington, R.S., Astron. J., 82, 753 (1977).

[30] Heggie, D.C., Mon. Not. R. astr. Soc., 173, 729 (1975).

[31] Heggie, D.C., "The Few Body Problem", IAU Coll. No. 96, ed. M.J. Valtonen, Dordrecht, Kluwer, p. 213 (1988).

[32] Heggie, D.C. and Rasio, F.A., Mon. Not. R. astr. Soc., 282, 1064(1996).

[33] Henon, M., "Families of Periodic orbits in the Three body problem," Astron. J.10, 375 (1974).

[34] Hills, J.G., AJ, 80, 809 (1975).

[35] Hills, J.G., AJ, 102, 704 (1991).

[36] Hills, J.G. and Fullerton, L. W., AJ, 85, 1281 (1980).

[37] Huang, T.Y. and Innanen, K.A., Astron. J., 88, 1064 (1983).

[38] Huang, T.Y. and Valtonen, M.J., Mon. Not. R. astr. Soc, 229, 333 (1987).

[39] Hut, P., ApJ, 272, L29 (1983).

[40] Hut P, The Big Bang and Georges Lemaitre, Proc, Symp. Louvain-la-Neuve, Belgium, Oct. 1983, ed (1984).

[41] Hut, P. and Paczynski, B., ApJ, 284, 675 (1984).

[42] Hut, P., Mc Millan, S., Goodman, J., Mateo, M., Phinney, S., Pryor, T., Richer, H., Verbunt, F. and Weinberg, M., PASP, 104, 981 (1992).

[43] Hut, P. and Tremaine, S., Astron. J. 90, 1548 (1985).

[44] Jakubik, M. and Neslusan, L., Contrib. Astron. Obs. Skalnate Pleso 39, 85 (2009).

[45] Jefferys, W. H. and Moser, J., Quasi-periodic solutions for the three-body problem, Astron. J., 71,568 (1966).

[46] Kaib, N.A. and Quinn, T., Icarus, 197, 221 (2008).

[47] Kroupa, P., Aarseth, S. and Hurley, J., Mon. Not. R. astrn. Soc. 321, 699 (2001).

[48] Leto, G., Jakubik, M., Paulech, T., Neslusan, L. and Dybczynski, P.A., Mon. Not. R. astrn. Soc. 391, 1350 (2008).

[49] Liu, X., Baoyin,H. and Ma, X., Astrophys. Space Sci., 339,295 (2012).

[50] Lovett, E. O., "Generalizations of the problem of several bodies, its inversion, and an introductory account of recent progress in its solution", Quart. J. Math., 42, 252 (1911)".

[51] Mazeeva, O.A., Solar System Res. 38, 325 (2004).

[52] Mc Cuskey, SW., "Introduction to Celestial Mechanics", Addison-Wesley (1963).

[53] Meeus, J., " Astronomical Tables of Sun, Moon and Planets", Willmann-Bell, Inc, Virginia (1995).

[54] Merman, G. A., "Qualitative research in the problem of three bodies," Bull.Inst. Theoret. Astron., Leningrad, 6, 687 (1958)".

[55] Mikkola, S. and Valtonen, M.J., Mon. Not. R. astr. Soc., 223, 269 (1986).

[56] Mikkola, S. and Valtonen, M.J., Mon. Not. R. astr. Soc., 259, 115 (1992).

[57] Moser, J, On the theory of quasi-periodic motion, SIAM Rev., 8,145 (1966).

[58] Moulton, F.R., " An Introduction to Celestial Mechanics", Ch. IX, Dover pub. Inc., New York (1970).

[59] Napier, W.M. and Staniucha, M., Mon. Not. R. astrn. Soc. 198, 723 (1982).

[60] Neslusan, L. and Jakubik, M., Astron. Astrophys. 437, 1093 (2005).

[61] Phinney, E.S., Phil. Trans. R. Soc. Lond. A, 341, 39 (1992).

[62] Phinney, E.S. and Kulkarni, S.R., ARA&A, 32, 591 (1994).

[63] Plummer, H.C., Mon. Not. R. astrn. Soc. 71, 460 (1911).

[64] Plummer, H.C., "An Introductory Treatise on Dynamical Astronomy". Cambridge Univ. Press, Cambridge (1918).

[65] Poincare, H., "Sur les groupes des equations lineaires," Acta Math., 4, 201 (1884).

[66] Prado, A. F. B. A. (2003), J. Guid. Control Dyn., 26, 33.

[67] Rasio, F.A. and Heggie, D.C., ApJ, 445, L133 (1995).

[68] Rosolowsky, E., Astrophys. J. 654, 240 (2007).

[69] Roy, A.E., Orbital Motion, Adam Hilger Ltd, Bristol (1978).

[70] Saslaw, W.C., Valtonen, M.J. and Aarseth, S.J., Astrophys. J., 190, 253 (1974).

[71] Shu, F.H., Astrophys. J. 214, 488 (1977).

[72] Siegel, C. L. and Moser, J. K., Lectures on Celestial Mechanics, Springer- Verlag , Berlin, p.138 (1971).

[73] Solomon, P.M., Rivolo, A.R. and Barrett, J., Yahil., Astrophys. J. 319, 730 (1987).

[74] Spitzer, L., Jr, "Dynamical Evolution of Globular Clusters". Princeton Univ. Press, Princeton (1987).

[75] Stromgren, E., “A new class of periodic solutions in the general problem of three bodies, viz. in the problem of three bodies with all masses and all distances of the same order of magnitude”, Publicationer og mindre Meddlelser fra Kobenhavns Observatorium,No.34(1919); also Monthly Notices Roy.Astron.Soc., 80,12 (1919).

[76] Sundaman, K. F., “Memoire sur le problems des trios corps,” Acta Math., 36, 105 (1912).

[77] Szebehely, V., “Solution of the restricted problem of three bodies by power series,” Astron. J., 71, 968 (1966).

[78] Szebehely, V. and Peters, C.F., Astron. J., 72, 876 (1967)

[79] Szebehely, V. and Peters, C. F., Astron. J. 72, 1187 (1967).

[80] Szebehely, V., Astron. J., 77, 169 (1972).

[81] Valtonen, M.J., Mem, R. astr. Soc., 80, 61 (1975).

[82] Verbunt, F., ARA&A, 31, 93 (1993).

[83] Whitworth, A., Summers, D., Mon. Not. R. astrn. Soc. 214, 1 (1985).

[84] Williams, J.P., Blitz, L., McKee, C.F., Protostars and Planets IV, eds., V. Mannings, A.P. Boss and S.S. Russell, Univ. Arozona Press, Tucson, 97 (2000).

[85] Wintner, A., “The Analytical Foundations of Celestial Mechanics”, Princeton, N.J.:Princeton University Press, (1941).

7.3 Publications

(1). Study of Motion of Three Bodies with Numerical Results

$K.Ramesh^1$ and $S.N.Hasan^2$

1. Department of Mathematics, Osmania University, Hyderabad, India.

2. Department of Astronomy, Osmania University, Hyderabad, India.

Global Journal of Theoretical and AppliedMathematics Sciences.

ISSN 2248-9916 Volume 1, Number 3 (2011), pp. 171-181

Abstract

We study the Three - Body - Scattering problem, first introduced by Burrau in 1913, where three bodies start from rest at the vertices of a Pythagorean right angled triangle. The bodies are assumed to be point masses and the Newtonian equations of motions are applied at close encounters. We perform simulation and investigate the nature of the solution for large intervals of time. We find that initially the three bodies interact forming double or triple systems which are short lived and ultimately the larger two bodies form a stable binary while the smallest body escapes the system.

Keywords: Three - Body - Problem, computer simulations, escape.

(2). Periodic Changes in the Synodic Period of the Moon and Its Relation with the Advance of its Perigee

K.Ramesh[1] *G.M.Ballabh*[2] and *S.N.Hasan*[3]

1. Department of Mathematics, Osmania University, Hyderabad, India.

2,3. Department of Astronomy, Osmania University, Hyderabad, India.

International Journal of MathematicalSciences and Engineering Aplications.

ISSN 0973-9424 Volume 7, Number IV (2013).

Abstract

As the regression of the nodes of the Moon's orbit is related with the occurrence of eclipse, the periodic changes in the synodic period of the Moon has been found to be related with the advance of the perigee of the Moon's orbit. From the variation of the synodic periods over 630 synodic months it is clear that there are two sinusoidal fluctuations in the synodic period. A larger sinusoidal fluctuation curve of about 112 synodic months and within this curve there are eight smaller sinusoidal fluctuation curves of about 14 synodic months. The dependence of the amplitude of variation with respect to various angular separations can be understood by the fact that the tidal effect on the Moon varies with its relative distance from the Sun, which in turn causes periodic changes in semi-major axis and the eccentricity of the Moon's orbit.

Keywords: Sun-Earth-Moon system, perturbation, advance of perigee, regression of nodes.

(3). Scattering phenomenon in binary systems

K.Ramesh[1] and *S.N.Hasan*[2]

1. Department of Mathematics, Osmania University, Hyderabad, India.

2. Department of Astronomy, Osmania University, Hyderabad, India.

Communicated in :*AstroPhysics and Space Science.*

Abstract

Huang and Valtonen (1987) obtained an analytic expression for the energy transfer in a parabolic encounter of a single star with a circular binary keeping terms of the first order in a/q where a is the radius of the circular orbit and q is the distance of closest approach of the single star. Heggie et al (1996) derived analytical expressions for the change in the orbital eccentricity of a binary following a distant encounter with a third star on a hyperbolic or parabolic orbit. We extend the Huang and Valtonen analytic treatment to obtain angular momentum transfer and change in eccentricity and illustrated the results numerically for different phase angles for direct and retrograde cases. Also we given detailed comparisons Heggie et al (1996) with the results of direct numerical integrations of the equations of motion for the three bodies. We considered for the case of a binary with zero initial eccentricity for different inclinations $i = 0°, 45°, 90°, 180°$. Also we studied numerically for the case of a binary with non zero initial eccentricities $e = 0.1, 0.01, 0.001$ and for different mass values of the perturber $m_3 = 1.1, 1.5, 2$.

Keywords: Celestial mechanics, stellar dynamics, binaries, changes in orbital elements.

(4). The Scattering of Oort's cloud objects due to encounter with Giant Molecular Clouds

K.Ramesh[1] and *S.N.Hasan*[2]

1. Department of Mathematics, Osmania University, Hyderabad, India.

2. Department of Astronomy, Osmania University, Hyderabad, India.

Communicated in: *Bull. Astr. Soc. India.*

Abstract

We estimate the effect of perturbation of giant molecular clouds (GMC) on the Oort cloud (OC) and to study the change of OC structure. In the present study, we solve the equations of motion of all considered test particles (TPs) using the FORTRAN code provided by Jakubik and Neslusan (2009). We model the OC as a Plummer sphere with 10^5 TPs and perform simulations and study the effects on the orbits of TPs at varying distance from the Sun in terms of semi-major axis a = 37000AU, 47000AU and 53000AU with varying inclinations i with respect to the ecliptic. We compared the path of the TPs with distance of semi-major axis a = 37000AU and a = 47000AU with inclination i = 0. Similarly change in semi-major axis a and change in q with respect to time for the TPs for initial semi-major axis a = 37000AU.

Keywords: Oort-cloud, giant molecular clouds, Plummer model.

More Books!

I want morebooks!

Buy your books fast and straightforward online - at one of world's fastest growing online book stores! Environmentally sound due to Print-on-Demand technologies.

Buy your books online at

www.morebooks.shop

Kaufen Sie Ihre Bücher schnell und unkompliziert online – auf einer der am schnellsten wachsenden Buchhandelsplattformen weltweit! Dank Print-On-Demand umwelt- und ressourcenschonend produzi ert.

Bücher schneller online kaufen

www.morebooks.shop

info@omniscriptum.com
www.omniscriptum.com

OMNIScriptum

FSC
www.fsc.org
MIX
Papier | Fördert gute Waldnutzung
FSC® C083411

Kolibri 360 GmbH
Ferdinand-Jühlke-Straße 7
99095 Erfurt, Deutschland
produktsicherheit@kolibri360.de

Druck:
CPI Druckdienstleistungen GmbH
im Auftrag der
Kolibri 360 GmbH
Ferdinand-Jühlke-Str. 7
99095 Erfurt